METAPHORMS AND INFORMATION PHYSICS

METAPHORMS AND INFORMATION PHYSICS

Louis M. Houston

AuthorHouse™
1663 Liberty Drive
Bloomington, IN 47403
www.authorhouse.com
Phone: 1-800-839-8640

First published by AuthorHouse   09/30/2011

ISBN: 978-1-4634-3324-6 (sc)
ISBN: 978-1-4634-3323-9 (hc)
ISBN: 978-1-4634-3322-2 (ebk)

Library of Congress Control Number: 2011912529

Printed in the United States of America

CONTENTS

Acknowledments

This book owes its existence to the work of Todd Siler and his concept of metaphorms. I am also in debt to Ken Wilber and his concept of consciousness and an integral vision.

There are also many personal interactions which greatly contributed to this book. Those were with Shawn Guillary, Gary Glass, Matthew Titsworth, Josh Sonnier, David Dawes, and Alexander Dymnikov. I am also indebted to certain family members which include Gwendolyn Houston, Gwen Houston, and Dominique Lueckenhoff.

Preface

This book bases its presentation on the concept of the metaphorm, presented by Todd Siler in his work on neurocosmology. A metaphorm is an underlying "super-concept" which creates commonality for distinct formats. An example of a metaphorm is the underlying concept shared by pressure and voltage. Both quantities stimulate the flow of a material. Pressure stimulates the flow of fluid, while voltage stimulates the flow of electrical charge. The approach to this document which is both speculation and science is that we have available to us two contrasting thought mechanisms represented by the right brain which is responsible for the artistic, philosophical and inductive and the left brain which is responsible for the scientific, logical and reductive. In order for this presentation to be complete, it was considered necessary to present both the philosophical and scientific sides of the themes presented. The essential argument for the metaphorm as I see it is that we can benefit from a synergistic combination of dual processing. This is also consistent with the original ideas of Todd Siler. This presentation promotes the consolidation of information. Besides philosophical arguments which aim to simplify our understanding of information, I present theoretical arguments which metaphorm electricity and magnetism into information rate or communication. The essential motive relies on the philosophical concept of dualism, the fundamental existence and contrast of opposites. The essential result is an equivalence between information rate and energy. To some extent this work should be considered pure speculation. Alternatively there is true theoretical science in this presentation. This combination of speculation and theory was considered to be a necessary presentation theme. It is hoped by the author that this work can induce the further exploration of metaphorms, particularly for scientific applications. In addition it is hoped that new approaches to science which also incorporate art, such as in this work, are encouraged. The scientific results in this work have only been subject to the tests of consistency presented. Consequently, they should only be taken at face value. Considering the novelty of some of the concepts presented in this work, there is a thin line between the designation of fiction and non-fiction. The fictional aspects of this work are more reflective of the speculative component, while the non-fictional aspects are more reflective of the science component. We consider this work as a story which is believable to the extent that it might be considered to be true. From this perspective we see truth as relative to the intensity and validity of the arguments made in support of it. Thus, we suggest that you suppress skepticism when reading this, and consider how valid the arguments seem in their totality. As a final point, we note that many of the ideas presented appear redundantly throughout the book. This was done intentionally to add reinforcement to ideas.

The Problem

It is one thing to understand what is better or what is necessary, but it is another thing entirely for one to achieve these things. Knowing what to do doesn't mean that one is capable of doing it. That's the problem in life. It's not that we don't know right from wrong. It's that we are often incapable of doing what is right.

The most difficult thing in life is to find consistent meaning and motivation. The wisdom in different areas of life is available to us, but the ability to follow that wisdom often eludes us. Whether it is the Bible, the Kaballah, Buddhism, Hinduism or Islam, the truth of how to lead a better life is available to us. The challenge is to find the will to follow the truth. The desire may be there, but we may simply have the inability to do it. That is something that many people misunderstand. They think that the problem with the world is misunderstanding, but that isn't it. The problem is will. We understand the right thing to do, but we have to have the will to do it. It is hard to say how to find the will if we don't have it. It is a problem that people have faced for eons. One of history's most famous philosophers, Nietzsche believed that the fundamental element was the will and that only certain few had the will to dominate. This all sounds very evolutionary as in survival of the fittest. You can lead a horse to water, but you can't force it to drink. The truth is that the horse may be thirsty, but not have the will to drink. In today's age of information, everyone has the truth available to them, but not everyone has the will to follow it. For example, Christianity says to turn the other cheek. We don't intrinsically know what this means and often we don't have the will to do it. We have to condition ourselves to have the will. With time and repetition information becomes more concrete. First the idea is abstract, then it eventually becomes a reality. We must constantly meditate on an idea in order for it to become a reality. Thus, we must first find a truth, then we must study it. We must observe the truth in many different perspectives over a long period of time and eventually the truth will manifest through us. It is the idea that tension leads to continuity which manifests a thought into reality. We must first build the tension between the thought and reality or act as if the thought is real. Nature will eventually create a bridge of continuity between the thought and reality (we will discuss this further later). A real connection will manifest to make the thought a reality. This is how we embrace truth. This is how we become a better person. We must first imagine ourselves to be better and then do the things that an improved version of ourselves would do. This creates a tension. Eventually we will become the imagined version of ourselves. This is where tension creates continuity. The energy in information is a measure of the energy required to learn the information. We invest energy in information in order to learn it. Learning is the process which transforms a thought into reality. Simple things require a minimal amount of learning. Complex things require a maximal amount of learning.

As human beings, we are challenged to live our lives in the midst of struggle, competition and misunderstanding. As time goes on we are confronted with more and more technological growth which is meant to ease life's burdens but also serves to overwhelm us with more and more information. In fact, the common denominator of all growing disciplines is information. The challenge to sort out and manage this information can create undo pressures and result in suffering. This ever increasing information consists of themes and anti-themes which often result in conflict within our inner mind space. What we say is that there is an outer universe and an inner universe. The inner universe is comprised of the self and all of its modes. A consistent and fluid interaction between the inner and outer universe is necessary to ensure a smooth and peaceful flow of information. The search for inner peace has gone on for millennia and has been facilitated by disciplines like philosophy, psychology and spirituality. Unifying themes meant to improve our understanding of the outer universe have persisted in fields like physics for many years. Religion has probably been the main proponent of inner peace. A major theme in Western Religion has been Christianity which points out the problem of sin and the solution of love and sacrifice. In contrast, a major theme in Eastern religion has been Buddhism which points out the problem of suffering and the solution of non-attachment and compassion. While religion has pointed to some form of spiritual growth as the solution to psychological affliction, we can clearly state that the fundamental problem is due to the clustering of information within the human psyche. As long as information continues to grow at more and more phenomenal rates, human beings will continue to be bombarded with conflicting and disorganized concepts which only serve to bring chaos to our inner universe and subsequently to the outer universe we interact with. Recently there have been attempts to unify disciplines under a form of consilience. Ken Wilber has developed what he calls an Integral Vision which categorizes and structures different disciplines within a common framework. Wilber also makes analogies to physics, but doesn't normalize descriptions to physical descriptions. What is proposed here is that physics works to explain the behavior of nature. However, physics is excluded from disciplines which do not focus on space and energy. Through the use of the metaphorm as described by Todd Siler, physics can be projected into other disciplines and consequently, the unifying principles of physics can also be projected into other disciplines.

Philosophy

The Buddha discovered that suffering is the primary problem in the world.

People are often poor and starve. Everyone gets sick and dies. There are also the numerous psychological discomforts. Life is a struggle and people often lose. He found that the primary cause of suffering is attachment. People cling to things that inevitably change because everything is impermanent. When the things we cling to change we suffer. In short, desire leads to suffering. By releasing all desire and attachment we achieve inner peace or nirvana. In relinquishing desire we also relinquish its complement, opposition. This suggests two things: don't desire and don't oppose. Buddha designed a system for achieving the desire-less state called the eight-fold path. The most important element of this system is meditation. Meditation is many things, but the primary impetus of meditation is relaxation. Meditation consists of achieving a mental focus on a simple thought pattern. It can be done by assuming a posture of stasis and observing our thoughts, while paying attention to something simple and repetitive like our breathing or a chant. Meditation calms the mind and trains it to not cling to thoughts. When thoughts occur during meditation, we simply observe them and let them be without desire or opposition. I have found that listening to relaxing music is very meditative. In time, meditation trains us to simply observe thoughts so that when we encounter real situations we are able to function without attachment. Some people think that it is implausible to live without desire because they feel that there would no longer be any motivation to do anything. The answer is simple. When faced with a decision, we can be motivated to do the right thing. There need be no desire involved in the process. That is why the eight-fold path includes the right ways to do things such as right effort, right mindfulness and so forth. Buddhism emphasizes compassion as a motivating factor in our interaction with the world. We should have compassion for others and be motivated to cause them no harm. Buddhism also has at its core the fundamental idea of duality. According to Buddhism everything is basically interdependent. There is no such thing as an isolated entity. This follows from the contrast between hot and cold or light and dark or male and female. We need one compliment in order to understand the other. Thus, at its core, reality is dual. This is consistent with the philosophy of the Tao, in which everything is a consequence of the interaction between yin and yang forces. Yin is the more passive, female force while yang is the more aggressive male force. Yin contains yang and yang contains yin, so that they are inter-related. There is a dynamic attraction between yin and yang which gives rise to all of the experiences in life. This is consistent with the Buddhist's rule of interdependence. According to Buddhism, everything has a cause. Because everything has a cause, we can account for every action in the world if we simply follow it to its source. This leads us to the concept of karma which is action. Since every action is caused and to each action there is a reaction, karma always points us to its cause. According to karma, when we do good things, good things will happen to us and when we do bad things, bad things will happen to us. Karma is consistent with the notion of thoughts creating reality. If we think positive thoughts, then positive things happen. If we think negative thoughts, then negative things happen. We can be encouraged by this because according to karma, false notions will negate themselves because of inconsistencies while true notions will reinforce themselves. As a consequence, we will never be forced to follow a false notion forever. This is the secret of enlightenment. As human beings following Buddhism we always have the potential to find inner peace. It doesn't matter what we follow, as long as it's true, it will eventually reinforce itself. If we are Christian and we believe that our faith is true, then it will eventually lead us to salvation. If we are Islam and we believe our faith is true, then it too will eventually lead us to salvation. We can confidently conclude that evil will never lead to salvation or enlightenment. Evil reduces itself, eventually. Good raises itself, eventually. The differences between Buddhism, Christianity, Islam or Hinduism are matters of focus. Buddhism focuses on logic while Christianity focuses on faith. Buddhism focuses on compassion while Christianity focuses on love. Buddhism sees emptiness or the Void as the ultimate truth. Christianity sees God or Christ as the ultimate truth. Buddhism believes that everything is ultimately empty because its existence depends on complimentary things. In an of itself, a thing is empty. There is evidence in modern physics that fundamental systems like atoms are basically empty, mostly occupied by space. Buddhism sees emptiness as an openness to possibility. The idea of emptiness is often misunderstood and can lead to nihilism or the view that nothingness is at the core of reality. If there is essentially nothing, then there is essentially no meaning. Suicidal thoughts have resulted from this type of philosophy. Buddhism believes that because reality is empty we have the opportunity to insert meaning where it is appropriate. Mystics believe that God is meaning. If you don't have a God then you must discover one.

There is a dual to the concept of no desire and no opposition. That dual is addiction. Ultimately, addiction is the cause of suffering. When we are addicted to something, we never get enough or we are never satisfied. After we use something that we are addicted to we require more and more of it to reach some semblance of satisfaction. Addiction is normally associated with drug use but it also applies to other bad habits, like over gambling or too much sex.

Duality is the concept that reality consists of themes and anti-themes and that a theme has meaning only in the space of the anti-theme and vice versa. For example, hot has meaning in the space of cold and cold has meaning in the space of hot or energy has meaning in the space of time and time has meaning in the space of energy. The ultimate compression of information results from the conclusion that duality is monism. That is, dual poles of meaning have in fact the same meaning. The reason why it is difficult to see duality as a monism is because we are only able to have one thought about one extreme at a time. If we were able to have dual thoughts about both extremes at the same time, it would be clear that duality is monism. The metaphorm gives the same meaning to a process which occurs in different spaces. Consequently, a metaphorm is able to see compliments as the same and is a mechanism for visualizing that duality is monism. We are reminded of the famous Buddhist statement, "form is emptiness and emptiness is form." In terms of the concept of extension, a string is a duality. It has a beginning point and an ending point or two nodes and one edge. Using the terminology of nodes, edges, and faces, the string is a simple network. In this sense, the nodes of the string form the poles of duality or a dipole. In terms of the concept of extension, string theory is the metaphoric description of all fundamental meaning. When the string vibrates and produces energy, reality happens. String theory is considered to be the possible theory of everything. However in its current form it is restricted to physics. The projection of physics into other disciplines through the use of the metaphorm creates a broader platform for string theory. Its basic structure as a meta-dual system which generates other systems through dynamic interactivity or vibrations represents the core structure of reality. While we suggest the relationship of string theory to other aspects of reality our focus will be on the electromagnetic field. As the repository of positive and negative interactions via charge, the electromagnetic field is the fundamental reflector of duality. Consequently, electromagnetism and string theory are fundamental metaphorms.

Information is divided into two complimentary formats: identity and location. When we know the identity of a symbol, we do not know its location. That location is confused with the location of other identical symbols. When we know the location of a symbol, we do not know its identity. As an example, consider the reception of an unknown message. At the moment the message is first received, before it is examined, we know exactly where the message is, but we do not know what the message is. This state represents location information. After examining the message, we know exactly where the message is but we no longer truly know the location of the symbols which represent the message because they occur elsewhere. This state represents identity information. When we examine ourselves in stasis, this is meditation. In this state our information is in the form of location. Consequently, we find no self in this state. In contrast, when we are moving and dynamic, our information is in the form of identity. Consequently, we find self (or soul) in this state. The no self or location information state is a thought-free (intuition rich) state that has been associated with nirvana and enlightenment. In contrast the self or soulful identity information state is a thoughtful (logical) state.

Almost ninety percent of our mental activity is located in the unconscious brain. The residual ten percent is located in the conscious brain. We are aware with our conscious brain. However, most of our mental processing takes place in the unconscious brain.

As we are exposed to more and more information, the unconscious must regulate more and more information using rules either implicit to the information or learned explicitly through the conscious channels. While education and experience plays a large role in managing the flow of information in the unconscious, it is only through specific rules of thought that blockages in the flow of information can be avoided. Since a great deal of the universe is mysterious, it is only through the more esoteric disciplines that we can usually gather the necessary guidance for certain mental activity. Such disciplines as spirituality and psychology seek to aid us in managing subtle mental activity. The universe as we envision it consists of themes and anti-themes which create both positive and negative forces which interact constantly. The accumulation and release of energy results from the interaction of these forces. Consequently, the controlled interaction and flow of information is required to avoid undesired energetic results such as depression and irrational mental states. While the outer universe processes energy, the inner universe processes information. Metaphorically, information is proportional to energy. If we are able to control information internally, we are able to control energy externally. By combining and organizing fundamental concepts, we are able to program our brain to process compressed information. The compression of information reduces the net flow within the brain and increases the possibility of the assimilation of more information.

The most pragmatic method for shifting from a perspective of duality to a perspective of non duality is the middle path. That is, we choose the middle point between two extremes. The middle path is the path of balance. In what might be called a meta-dual perspective, we choose one or the other and simultaneously recognize both as being the same. The meta-dual perspective is not to be confused with the non dual perspective which selects both poles (extremes) as being the same. The meta-dual is similar to the dual in the sense that it selects one pole while recognizing that pole to be equivalent to the other. It was pointed out earlier that the inability to have two thoughts at once is what limits the experience of the meta-dual perspective. The meta-dual perspective is a form of monism in which what would be two simultaneous thoughts are differentiable yet equivalent. This is similar to the process of stereo in sound systems. Meta-duality is the ultimate form of compression since it is a monism. However, meta-duality maintains differentiation. In meta-duality the two poles are the same yet different. Metaphorically, in meta-duality we hear two sounds which are distinct, yet compliment each other. In the middle path, we here one sound which is the average of the two original sounds. In non-duality, we hear one sound. It is only if we limit the brain to have only one thought at a time that the experienced form of meta-duality lacks simultaneity as in stereo. Just as in music when two notes augment each other when they act together to form a chord, a theme and an anti-theme augment one another when they act together and go beyond the experience of either individually. We shall refer to a normal duality experience as "mono" and refer to a meta-duality experience as "stereo". The challenge is to develop the software needed to program the brain such that mono experiences get converted into stereo experiences. This achieves both compression and enhancement of the information.

Traditionally thought is separated from emotion. However in this theory, emotion is a form of thought. Thought is a form of information. Information is metaphorically equivalent to the electromagnetic field which has a spectral range. Just as light is part of the visible spectrum of electromagnetic radiation, symbolic information represents the "visible" portion of information. Consequently, in this metaphorm, information has wavelengths or frequencies. The lower frequencies correspond to instinctual information. The higher frequencies correspond to emotion. The even higher frequencies corresponds to irrational thought. When information is compressed, higher frequencies of thought are reduced to lower frequencies. Consequently, emotional expression can be reduced to symbolic expression. Similarly, symbolic expression can be reduced to instinctual expression. Just as there is an electromagnetic field, there is an information field. The information field is the source of concepts or ideas. As we are constantly exposed to the information field, we constantly extract ideas from it. In order to exploit the information field intellectually our brains must be able to transform the pure information into a symbolic form.

Exposure to drugs and the existence of chemical imbalances in the brain sometimes makes the brain more sensitive to the information field. The influx of too much information can cause mood swings and delusion. Programs which encourage the compression of information in the brain help to reduce the undo influence of an over abundance of information influx. When the brain short circuits on an information overload it can go into a state of depression. In depression there is very little contact with the information field. This lack of contact is the reason why many creative people are depressives because they are compelled to be creative in order to function. The ideal solution to problems of excess information radiation is to process information in stereo rather than mono.

From now on, when referring to themes and anti-themes, we will simply use the word "themes" to include both poles. As we have pointed out, normally, the brain can only experience one thought per unit time. Under normal conditions, in order for many thoughts to be experienced by the brain, the process of time-multiplexing takes place. In time-multiplexing there is some given cycle time or clock cycle and a succession of sub-intervals during which different themes are visited. For example, given a total of three themes, a thought from theme number one is experienced during the first sub-interval, then during the second sub-interval a thought from theme number two is experienced, then during the third sub-interval a thought from theme number three is experienced. The cycle then repeats itself and subsequent thoughts from each theme are time-multiplexed again. This multiplexing repeats until the complete themes are each visited by thought. In an actual brain there are many themes which are visited by the full spectrum of thought. Only the portion of the thought spectrum which is symbolic becomes part of the conscious brain. Simply put, the unconscious brain is the invisible portions of the thought spectrum visited by thought during time-multiplexing. Alternatively, the conscious brain is the visible portions of the thought spectrum visited by thought during time-multiplexing. Since there are thousands of themes in the brain, a clock has a very high frequency which implies that a sub-interval is a very small time interval. The gaps that we experience while thinking are time periods when invisible portions of the thought spectrum are occurring. Note that we use the term "thought spectrum" which has the same meaning here as "information spectrum." The structure of a theme ranges from simple to moderate to complex. A simple theme requires a small amount of information while a complex theme requires a large amount of information. This implies that in general, during time-multiplexed thought, there is irregularity to information flow as themes vary in their information requirements. We might think of the brain as a huge serial network which simulates parallel processing when the information flow is de-multiplexed. The flow of information through the brain is generally irregular due to the variation in complexity of different themes. This brain structure has an additional property. It has the ability to vary the length of a multiplexed sub-interval so that more or less time can be spent on different themes. This modulation of sub-intervals is equivalent to redundancy of a sub-interval. When we concentrate on a particular theme, we increase the redundancy of the sub-interval during which that particular theme is multiplexed, effectively increasing the length of the sub-interval. This implies that the brain has the ability to spend more time on some themes and less time on others. For any given individual, the level of concentration available determines what level of sub-interval redundancy can be incorporated. When an individual is in a state of relaxation or meditation, the brain is focusing on simple themes. When an individual is in a state of concentration, the brain is focusing on complex themes. The more complex the theme, the more energy dissipated by the information flow through the theme. When the brain is multiplexing thought such that there is a focus on complex themes, the energy dissipation can be very large. When we refer to energy dissipation in this context, we are referring to the total information flow processed per unit time. Sensitivity to the information field can stimulate concentration which in turn can cause large energy dissipation if left unchecked. Meditation can reduce the energy dissipation and counterbalance the effects of concentration. When the probabilities present in a theme are large or near unity, the information flow is small and so is the energy dissipation. When the probabilities present in a theme are small, the information flow is large and so is the energy dissipation. In other words, a highly probable thought has low information while an improbable thought has high information. A brain system which contains a network of highly probable thoughts contains a high level of causality and thus contains a high level of rationality. In contrast, a brain system which contains a network of highly improbable thoughts contains a low level of causality and may contain irrationality. The content of information can be identified by its frequency. Emotion is due to higher frequency information than that which causes symbolic thought. That is why efforts to concentrate will often lead to emotional outbursts. In short, emotional expression is a method of channeling "information power" to the body.

Science and Philosophy

It is true that information is proportional to energy and explicitly true that information is quantized. The smallest amount of information is a bit or the log base two of the number of possibilities to the answer to a yes or no question ($\log_2(2)=1$ bit). The primary study of information is derived from the discipline of Information Theory. Information theory, primarily developed by Claude Shannon intends the most accurate reproduction of a communicated symbol. Consequently, information theory proposed the format of information in terms of probability and studies its behavior under different conditions. The primary measure proposed by Shannon is called entropy, similar to thermodynamic disorder. Entropy measures the average a priori uncertainty carried by a message. It is given as $H = -k\sum_{i=1}^{n} p_i \log p_i$, where p_i is the probability that the ith event will occur. Entropy is measured in bits in which the logarithm is in base two and k=1. For example, if the event is the flipping of a coin then there is an equal probability of ½ that either outcome, heads or tails will occur. The total entropy is thus one bit. The behavior of information flow is metaphorically equivalent to the behavior of energy. When a blackbody absorbs and re-radiates electromagnetic energy, the density spectrum is peaked. (Prior to quantum mechanics and the discovery that energy is quantized, it was calculated that absorbed energy would be re-radiated at higher and higher frequencies asymptotically. This was referred to as the ultraviolet catastrophe.)

Consequently, there is a peak to the radiation of information which is absorbed from the information field. As the peak is passed, information will be re-radiated at lower and lower intensities as it reaches higher and higher frequencies. Consequently, a brain which is over-stimulated by the information field will eventually reach a peak in energy dissipation.

Time reduces information content. Repetition of the same thought reduces the information content of the thought. Consequently, meditation reduces information content. (We want to point out here that although we acknowledge that there are various forms of meditation, we will refer to meditation as the method most consistent with relaxation.) As thoughts are revisited and re-examined, the representation of information becomes more and more concrete. Thoughts, when carefully nurtured, become reality.

In other words, thoughts as information in the internal universe can become energy in the external universe. However, contradictions negate thought. Therefore if a thought is abstract and becoming more concrete but it encounters a contradiction with some idea which pre-exists in reality, then the thought will be negated.

The brain can be trained to meditate during periods of concentration when there is a need to reduce the information power. Meditation normally involves a focus on breathing or a chant while thoughts are observed and ignored. This normally occurs in a lotus sitting position or while the body is in stasis. I have discovered that music can be used to perform meditation by training the brain to play out thoughts of music in repetition. The brain focuses on the musical repetition and simulates a meditative state. The meditation helps to regulate the thinking by reducing the overall energy dissipation.

When an idea is new, it may be very abstract and highly imaginative. In time, with the proper evolution, the idea can become concrete and instinctual. It is important that the idea does not contradict any major themes in the outer universe (as stated earlier).

It occurs that identity and location information are correlated to kinetic and potential energy. It is also clear that due to the impact of realization, inertia is present in identity information, while potential energy is present in location information for reasons of complimentarity.

Knowledge separates the subject from the object. Before knowledge, identity is entangled. After knowledge, identity is disentangled. When we know something, we separate ourselves from it. Knowledge of a thing is the ability to deconstruct and re-create the thing. Knowledge is the ability to separate the foreground from the background. In this sense, knowledge of information compresses it and makes it more fluid. When information is unknown, it is consolidated. When information is known, it is distinct. Consequently, it is the ignorance of information which can cause blockages in its flow. Mysterious conglomerates of information produce mass blockages. These blockages naturally compartmentalize information so that flow is restricted to certain channels and blocked from others. Information blockages cause inefficient use of the brain. The acceptance of such conclusions as "life is a mystery" is not a viable solution to the problem of information management.

It is important to construct reasonable theories about themes. Truth is relative to the existing framework of information in the knowledge network being used. Consequently, absolute truth is a rare phenomenon. However, once a knowledge network is established, it is important to produce consistent networks that lead to no contradictions.

In the universe we live in, there is already an existing knowledge network. This is likely true in other universes. Each brain consists of its own unique inner knowledge network. The challenge is to build internal knowledge networks which are stable and capable of efficiently processing external information and creating energy representations in the external universe.

There are some questions about the relative functionality of an internal knowledge network. Our personal knowledge networks are shaped by our environments, our experiences and our heritages. The extent that we are able to compete with other knowledge networks does not determine our own personal consistency and stability. This is because our personal knowledge networks are unique and cannot be specifically compared to some other network. In other words intelligence is not easily normalized.

The only way to normalize intelligence is to compare the effective communication with the external universe. This can only be done relative to the existing external knowledge network. Our own internal stability is not a measure of intelligence, since stability depends on the structure and complexity of our internal knowledge network.

Intelligence is a measure of our ability to provide consistent explanations, not a measure of the size of our internal knowledge networks. However, the structure and complexity of our internal knowledge network might be indicative of what is called our "intelligence potential." By intelligence potential, we mean our equivalent intelligence if the internal and external knowledge networks were exchanged. In other words, our intelligence potential indicates how intelligent we would be considered if the rules for the management of information established internally were projected to the outside world and our intelligence were measured in comparison to others.

When I first started working on this book, I had to make spurious notes and then re-read them and think about them. As time progressed, I was able to make more detailed descriptions of the ideas within my notes in newer notes. I have finally progressed to writing the ideas for the book down directly. I think that this provides evidence that information management has evolved for me. It is evidence that the compression of information leads to better information flow in all themes. The presentation of the ideas in these writings lead to more and more concrete representations of the information themes. Subsequently, more ideas are able to flow.

Based on the words of Bacon, it is commonly believed that knowledge is power. Power is energy per unit time. Knowledge is information collected over time. Consequently, based on this premise, information is energy. The idea that knowledge is power drives the notion that there is tension in continuity. We propose that tension creates continuity is the rule that causes the capillary force to move a fluid against the pull of gravity in a siphoning process. The more information flows, the better it flows. Irregularities in the flow of information point to small blockages which must amount to small inconsistencies. There is a certain current level of inconsistency with using metaphor interchangeably with fact. For example, we will discover that there is a certain fuzziness to the definitions of information and information flow. These inconsistencies reduce as the level of abstraction of the information reduces with time and repetition. In other words, as we discuss new ideas and use them, the new ideas facilitate the flow of even newer ideas which connect in some causal fashion to the original ideas.

The recreation of a thought experience in the internal universe as an energy experience in the external universe is an extension of the notion of causality. Causality might be defined as the meaningful sequencing of events. When two sequential events are causally related, there is actually no force which connects them. This was pointed out by Hume.

Consequently, although we see causality as a real force, it is actually fictitious. Causality might be described as a compression of coincidence or when coincidental events are seen as meaningful. While not real in the physical sense, it is clear that causality is real on some level and is able to directly affect our physical experience. I posit that when a thought experience seems to become a reality, causality is present, making the juxtaposition of the two events meaningful. For example, if I think about acquiring money and then subsequently, I acquire money with no connective action, then these two events seem to be causally related although there was nothing that I specifically did to enable a connection between the two events. If these types of situations repeat themselves, then there would apparently be compressions of coincidence occurring.

The idea here is to first think about something and then act as if it has already happened and then causality will fill in the blanks. Once again, the principle here is tension creates continuity. Acting as if a thought has created an action supplies the tension. Continuity supplies the connective action needed to make the events meaningful.

Consciousness is awareness. In that sense consciousness is like a mirror. If there is no light, there is no reflection, or if there is no thought, there is no consciousness. We see ourselves in a mirror. We see ourselves in consciousness. Since symbols represent the visible spectrum of information, it makes sense that symbolic thought is the form that we are aware of. When we think and listen to our thoughts, our thoughts occur symbolically as patterns of language. Consciousness reflects information in two different ways: it can focus information as in a concave mirror or it can diffuse information as in a convex mirror. In the way that consciousness acts as a concave mirror, it involves concentration. In the way that consciousness acts as a convex mirror, it involves meditation. The process of evolving a consciousness is like polishing something down to a smooth surface. The smoother the surface, the more readily it reflects. As clarity in consciousness increases, the quality of thought increases.

When we talk about a unifying theory, we are trying to discuss a commonality or a causal link between apparently different themes. We are discussing the unification of forces. First of all, these forces all belong to the same universe. Second of all, these forces are all ways to store or release energy. Third, these forces are all interactions within a yang and a yin dipole. There are four forces (or three if two are combined), gravity, electromagnetism, the strong force, and the weak force. Gravity is the interaction between mass and space. Electromagnetism is the interaction between the positive and the negative. The strong force is the interaction between the strong and the weak. The weak force is the interaction between determinism and randomness.

Information is relative to motion. This motional behavior is like a Doppler shift. Moving information has a shifted frequency. Much like the sound of a train whistle, when information is moving towards you, it has a higher frequency. When information is moving away from you, it has a lower frequency. Consequently, it is easier to understand transmitted information than it is to understand received information or it is easier to understand what you write than to understand what you read. That is why most people tend to talk more than to listen.

The concept of interference seems to be important. When two thought trains contain contradictory information, the thought trains will interfere destructively. When two thought trains contain consistent information, the thought trains will interfere constructively. Similar to the interaction of matter and anti-matter, a theory is destroyed when it meets an anti-theory. The theory is reinforced when it meets a correlating theory. As mentioned earlier, it is important not to have contradictions when constructing paradigms for the manipulation of themes. In particular it is important not to have contradictions between the inner and outer universes, as information flow is negated when contradictions are encountered. This is metaphorically equivalent to destructive interference between waves.

The Tao is a system which incorporates the interaction between yin and yang forces. The forces flow and merge into one another in unending cycles. The universe is driven by the Tao. The yin is the more feminine, more passive and negative force while the yang is the more male, more aggressive and positive force. One can visualize the ebb and flow of yin and yang forces within information as one can also see this within the flow of energy. The shield is yin. The sword is yang. The bow is yin. The arrow is yang. Block (yin), thrust (yang), block (yin), thrust (yang) is a fighting sequence. Yang is localized. Yin is global. Yang is particle. Yin is wave. Yin is defined in the space of yang and yang is defined in the space of yin. Mania is a buildup of yang. Depression is a buildup of yin. A buildup of yang is explosive. A buildup of yin is implosive. Yang is freedom. Yin is safety. The polarity of the information is determined by its yin and yang content. The amplitude of the information is the Shannon entropy. Shannon entropy is a measure of the average amount of surprise in the information. Explicit is yang. Implicit is yin. Hot is yang. Cold is yin. Strong is yang. Weak is yin. Deterministic is yang. Randomness is yin. Mass is yang. Space is yin. Positive is yang. Negative is yin. Consequently, the unifying forces of the universe are ordered into a yin and yang structure. In fact, all of the interacting concepts can be broken down into yin and yang pairs. Oscillations of yin and yang determine the frequency of information. Consequently, yin and yang oscillate faster in emotion than in symbolic thought and yin and yang oscillate slower in instinct than in symbolic thought. Yang flows into yin and releases energy. Yin flows into yang and absorbs energy.

Science and Metaphorm

It is of interest to discuss the equivalence of energy and information (rate). This equivalence is much like the equivalence of energy and mass in that there is a proportionality constant. We introduce the concept of "learning" which is the energy per unit information. Learning=energy/information. Thus, information= energy/learning. When the energy is small and the information is large, then learning is small. When the energy is large and the information is small, then learning is large. An example of the latter is that of rote learning. An example of the former is a one time encounter with a complex subject. Since information is relative to the observer, then learning is relative to the observer.

To the extent that certain thoughts can seed the decay of meaning, information is radioactive. Meaning is only maintained with consistency. When certain thoughts interfere destructively with the meaning of other thoughts, the result is contradiction and negation. In other words, when a thought with meaning interferes destructively with another thought with meaning, the meaning of each thought is destroyed. On the other hand, when a thought with meaning interferes constructively with another thought with meaning, the meaning of each thought is augmented. Without meaning, information doesn't flow. Meaning is the motive force behind information flow. While information flow is current, meaning is voltage. When voltage is low, current is low. When there is zero voltage, there is zero current. Metaphorically, voltage is pressure, so meaning is pressure.

Pressured thought is driven by meaning. When there is zero pressure there is no fluid flow. Consequently, as stated earlier, when there is no meaning, there is no information flow. From this point we can examine the meaning of resistance. The greater the resistance, the less the information flow. The lesser the resistance, the greater the information flow. The resistance presented by an event is a measure of the redundancy of the event. The more redundant the event, the greater the resistance to information flow. The less redundant the event, the less the resistance to information flow. In this sense, resistance is a measure of probability and conductance is a measure of inverse probability (conductance=1/resistance). We might say that information flows through an event to the extent that an event carries information. The more unique the event, the more information it carries. The more redundant the event, the less information it carries. Traditionally, the information carried by an event is measured as the log of one over the probability of an event or the negative log of the probability of an event. In a sense, we can say that information is the log of the conductance. This is called the self-information. Just as voltage is the sum of current*resistance (IR) drops across a circuit, meaning is the sum of information*probability drops across a circuit. We make one modification to this process. The equivalent resistance used to multiply the self-information is the conditional probability. This modifies the Shannon entropy to be weighted with conditional probability rather than probability itself. The traditional Shannon entropy can be written as

$$H = -\sum_{j} p_j \log p_j$$ while the modified entropy which measures

meaning is given as

$$M = -\sum_{j} p_{j|j-1} \log p_j$$ in which we observe the conditional

probability as the weight. Naturally, the more coherent a sequence of events, the larger the meaning. We point out here that the interval of the conditional probability determines the scale of the meaning. Since we have specified a nearest-neighbor interval, the corresponding meaning is at the first scale.

From the point of view of information, this modified Shannon entropy or meaning (M) is the sum of IR drops across a circuit of thoughts. As pointed out, metaphorically, the modified entropy is meaning. Using this analogy to circuits we can examine a thought that consists of either a "yes" or a "no".

The current flow of information through the thought is one "amp". Allowing the conditional probabilities to be ½, the meaning or voltage across the thought is one half "volt". Using Joule heating which is equal to I^2R, we can determine the energy dissipated in terms of power by the thought as one half Watt. In a time-multiplexed circuit with a sub-interval of one second, the energy of the thought would be one half Joule.

The fundamental issues with thoughts manifesting themselves into concrete reality are first, thoughts can be and are made unconsciously. Second, inconsistent thinking produces destructive thoughts. As a culture, we are inadvertently destroying ourselves. We wouldn't intentionally destroy ourselves but we have very little control over our brains. People believe that thoughts become reality to the extent that it must be true. The problem is that it is not just positive thoughts that become reality. From an evolutionary level, this power is self-destructive in the defense of the outer universe. When we misuse our thoughts, we suffer the consequences. The argument that thoughts only exist in the conscious mind is false, because thoughts can exist in different frequencies beyond the visible or symbolic spectrum. A unifying paradigm must encompass the full spectrum of thought and thus offers a solution to the problem of inconsistency in unconscious thought.

There is some resonance in the fact that yang reduces to yin and positive reduces to negative are correlated. Positive reduces to negative is the electromagnetic force interaction that is recognized because negative is associated with ground. Electromagnetism is the primary metaphorm for information. We see information as a wave comprised of yang and yin amplitudes, which are essentially positive and negative, respectively. With reference to electricity, we also metaphorm information flow into current and interpret voltage as meaning. Consequently, we see that information is primarily electromagnetic in form. This leads us naturally to interpret probability as resistance, using Ohm's law. It is interesting to note that while current seems to be the particle-like interpretation of information, electromagnetic radiation is the wave-like interpretation of information.

Referring to electrical circuits, we can metaphorm attachment into capacitance or the ability to store charge. As the yang pole is positively charged and the yin pole is negatively charged, we can interpret capacitance as the ability to store yang and yin polarity. The strength of yang and yin relate to information content or amplitude. A high level capacitor is able to store a lot of charge consisting of a large information content. A low level capacitor is able to store only a minimum of charge consisting of a low information content. In Buddhism the goal is non-attachment or in this sense, zero capacitance. The Buddhist tenet is consistent with lack of charge storage (i.e., information storage).

Electric force is the attraction between yin and yang. Electric current is the flow of yin or yang. Magnetic force is the attraction between different currents (i.e., information). Consequently, magnetic force occurs between different knowledge networks. The affinity that two individuals may have for one another is due to magnetic force. It follows that the brain which is a knowledge network produces magnetic fields. In the sense that the brain is also a true electrical network, it produces true magnetic fields. We can say that the metaphorical network and the true network merge with the manifestation of magnetic fields. From this perspective, we can say that consciousness is equivalent to magnetic flux. The earlier metaphorm of a mirror for consciousness is maintained by acknowledging that the magnetic force is a "mirror" of the electric force. The magnetic field is in this case the awareness or reflection of the electric field.

The property of induction occurs when one knowledge network induces information flow in another knowledge network. Inductance is a metaphorm which is proportional to learning. With this definition we can refine the relationship between information and energy. Recall earlier that we had defined learning as the ratio of energy to information (learning=energy/information). From electromagnetic theory, we know that the energy (U) stored in an inductor (L) is given as $U = \frac{1}{2}LI^2$, where I is the current (or information rate). Consequently, the original equation is modified to learning=2*energy/information rate^2. It is interesting to note that the symbol for current is already consistent with information rate (I) and the symbol for inductance is already consistent with learning (L).

The electric field is parallel to the current flow. Consequently, the direction of the electric-information field is from left to right or parallel to this sentence structure ($\rightarrow$). In order to visualize the direction of the magnetic-information field, visualize a current which circulates along the left-right direction or in this case the x-axis. The curl of that current creates a magnetic field normal to the left-right direction or in this case the y-axis. Consequently, the direction of the magnetic-information field is down to up or normal to the electric-information field ($\uparrow$). The symbol for the electric field is $\vec{E}$, while the symbol for the magnetic field is $\vec{B}$. The electromagnetic radiation is called the Poynting vector, $\vec{S}$.

The equation for the Poynting vector is $\vec{S} = \dfrac{1}{\mu_0}\vec{E} \times \vec{B}$, where μ_0 is the permeability or the ability of a material to sustain a magnetic field. The Poynting vector gives the direction and magnitude of radiation. In this case, for the information on this page we have a direction given by $\rightarrow \times \uparrow = \cdot$, or out of the page. This is consistent because it uniquely determines the direction of information radiation to be out of the page towards the observer (i.e., you). To the extent that the electric-information field and the magnetic-information field have fixed directions, information is polarized. The polarization of information implies, in the case of the English language, that it is only meaningful when it is propagating from left to right.

There is a resistance to the induction in a circuit which results in a reverse emf (i.e., electromotive force or voltage). This is called Lenz's law. This is equivalent to saying that there is a resistance to the change in an idea. Consequently, ideas have inertia. This result is consistent with Thomas Kuhn's ideas about scientific revolutions.

It is important to note that while there is a strong similarity between duality and the yin and the yang of the Tao, there is a significant difference. While duality consists of opposites, the yin and yang are not purely opposite. There is a little bit of yang within the yin and there is a little bit of yin within the yang. This inhomogenuity is represented by two small circles within the Tao symbol. Because we are metaphorming duality into electric charge, our symbol is the electric dipole. While the Tao is appropriately non-dual, our information dipole is meta-dual. Our system is a meta-duality because we are metaphorming information as duality into an electric dipole. The field lines of the dipole are part of the electric-information field. Just as electric current produces a magnetic field which circulates the current, information flow produces a magnetic-information field which circulates the information flow. The force between adjacent words is due to the electric-information field while the force between adjacent sentences is due to the magnetic-information field. Metaphorically, we say that the information force is an electromagnetic force. The electromagnetic force, which holds material bodies together, in this sense, also holds information together.

Meta-duality

It is very simple. Think good thoughts and you will do good things. Think bad thoughts and you will do bad things. Thoughts become reality. In other words, tension creates continuity and continuity creates tension. This is a symmetry. Thoughts create tension. When we think about something we create a tension between our imagination and reality. When the thought that we have imagined becomes a reality, there is a continuity between imagination and reality. Good and bad are extremes. When we practice Buddhism and follow the middle path between extremes our thoughts are no longer black or white, they are gray. Gray thoughts create a gray reality. This is why we say that reality is fundamentally empty in Buddhism. Emptiness is the in between state of two extremes. Emptiness is essentially the zero state. The only solution to emptiness is God. Emptiness is the yin while God is the yang. Emptiness is the background and God is the foreground.

We present here an alternative non-dual way. This is the way of meta-duality which simultaneously embraces both extremes as in stereo. By encompassing both extremes simultaneously, meta-duality goes beyond both and is consequently non-dual. This yields what we call the omega state. While emptiness is gray, the omega state is in color. It encompasses the full information spectrum. The ideal way to experience the omega state is with two simultaneous thoughts. It is within the realm of human potential to sustain one passive background thought simultaneously with one active foreground thought. This can be accomplished by using the left and right hemispheres together. The left hemisphere provides the active thought, while the right hemisphere provides the passive thought. The left thought focuses on one extreme, while the right thought focuses on the other extreme. There is a way to train the brain to learn to act meta-dual although it may occur naturally for some. Visualize the brain as an air conditioner. The environment of the brain supplies the thought energy to the left brain which sings a song to the right brain. By "sing" we mean that the left thought duplicates music played back in short cycles. This becomes a chant for the observation of the right brain, which will appropriately "cool" down, reducing its emotional output. In this way, the brain creates a background for awareness which consists of two extremes, the hot yang of the left brain and the cool yin of the right brain. As the mental environment dissipates energy, the left brain gets hotter while the right brain gets cooler. Consciousness or awareness then develops in color. At some point after much conditioning, it is possible that consciousness will awaken and develop the ability to think actively. Thus not only will consciousness listen, it will also speak. In this way, we develop tertiary thought. The left is positive one, the right is negative one, and consciousness is zero. At this point, thinking is particularly meta-dual. There are three simultaneous thoughts. It is nonetheless advantageous to develop stereo thought which is sufficient for meta-duality. The stereo basis forms an orthogonal pair with the left thought being orthogonal to the right thought. As an orthogonal basis they are capable of representing all possible thoughts, continuously.

In this mode of thinking we are able to experience extremes simultaneously such as pleasure and pain. The middle extent of such an experience is moderate pleasure mixed with moderate pain. In this way the omega state contrasts with emptiness. The omega state allows for the full spectrum of experience, while the middle way gives only a gray projection.

The extension of logic into a tertiary state which includes yes, no, and uncertainty as three separate logical states is a meta-dual extension. This suggests that the division of reality into components of three as opposed to a duality is a fundamental reorganization. This is reflected in the division of the fundamental forces of nature into three, the strong, the electroweak, and gravity. Thus, the fundamental forces form a tertiary system which resonates with tertiary logic. There are other similar divisions. For example time becomes past, present, and future, color becomes red, green, and blue, mass becomes solid, liquid, and gas. Further examples include rock, plant, and animal as one, black, gray, and white as another, hot, warm, and cold as another, mass, energy, and information as another, x, y, and z as another, instinct, emotion and logic as another, negative, zero, and positive as another, and low, medium, and high as another. It is interesting to note that quantum mechanics does not recognize time as an observable quantity, which may explain its lack of representation as a position coordinate. Alternatively, position could be represented by x, y and t, consisting of a two-dimensional space and time as a sufficient model of reality. Tertiary logic is consistent with quantum mechanics which includes uncertainty as a viable logical state. Quantum mechanics characterizes physical systems as having wavelike extensions until observed (or exposed to a reactive system). The presence of force fields or boundaries will allow only discrete energy states to exist within these wavelike extensions. Thus the name "quantum" occurs. Observation perturbs the quantum mechanical system into a particle-like state. This is called wavefunction collapse. Waveforms have complimentary or conjugate representations in position and frequency spaces. Resolution of waveforms varies approximately inversely in conjugate spaces. This creates an uncertainty principle. The uncertainty principle can also be derived from consideration of commutation between conjugate operators. The waveform which characterizes a quantum mechanical state is generally derived by solving the Schrodinger equation. Once the waveform, ψ, is found, it yields the probability distribution for the outcomes of different observations of the physical system by squaring the waveform (i.e., ψ^2). Quantum mechanics suggests a link between the purely imaginary and the real. We might say that a quantum mechanical state is imaginary until it is observed and subsequently brought into reality. This is consistent with the fact that the wavefunction is complex, consisting of a real part and an imaginary part. The measure of the quantum mechanical system is the modulus which is a positive number. One produces probability by squaring the wavefunction. The probability measure is consistent with potential existence of a state. There are three states in quantum mechanics. They are existence as the positive logical state or non existence as the negative logical state or uncertainty as the ambiguous logical state. The uncertainty state is consistent with an imaginary existence. Interpretations of quantum mechanics vary and are too numerous to discuss here. Suffice to say that quantum mechanics supports a tertiary interpretation of reality.

A metaphorm is a concept developed by Todd Siler. It is essentially a metaphor which indicates an underlying process that has commonality between apparently different systems. We might say that a metaphorm is a meta-format or different formats for the same concept. A fuzzy set consists of a set A and a membership function, $m \in [0,1]$. For a typical set, the logic states that a variable is either a member or not a member. Thus in the case of a typical set, the membership function would be $m \in \{0,1\}$. In contrast, fuzzy set membership has a range of values. We can describe a preliminary range of values for set membership with the metaphorm by comparing subject, action and object. In this case we would have $m \in \{0, \frac{1}{3}, \frac{2}{3}, 1\}$. We might imagine that for continuous divisions of a sentence other than subject, action, and object, we might get closer to a continuum of membership. In the sense that equivalence is not typically of a specific value for fuzzy sets, it follows that metaphorms are fuzzy sets. Given two sentences describing two different systems, the systems are metaphorically equivalent at three different levels. Level one incorporates one level of commonality. This is either common subject, common action, or common object. Level two incorporates two levels of commonality. This is either common subject and common action or common subject and common object or common action and common object. Level three incorporates three levels of commonality. This is common subject, common action, and common object. The essential metaphorm proposed by Todd Siler in his work on neurocosmology is given by the following two sentences.

Brain synthesizes thought.
and
Star synthesizes energy.

This can be viewed as a level two metaphorm to the extent that the actions are the same and thought is proportional to energy.

Another example of a level two metaphorm is the following

*voltage imparts ch*arg*e flow.*

and

Pr*essure imparts fluid flow.*

A simple example of a level three metaphorm is the following

Roses are red.

and

Violets are blue.

At this point we propose that a level three metaphorm is an identity between two systems.

The essential structure is subject, action, and object. If each has equivalence to another subject, action, and object, then the ensemble is equivalent to the other. Thus, we have an identity between the two representative systems.

We can apply this idea to the following metaphorm.

Information flow is words per unit time.

and

*Ch*arg*e flow is ch*arg*e per unit time.*

Using the result of duality which charges information, we see upon examination that these two statements refer to identical systems. This supports the essential proposition of this book.

Intelligence

Axiom I. Information is proportional to length.

As this work develops we will explore the consequences of axiom I.

Consider the communication process. Two communication systems exchange symbols at some given rate. The symbols have different probabilities of occurrence. The information transmitted and received is the self-information of each variable in the sequence communicated. We can transform the communication process into the apparently random selection of phase from synchronized clocks. If the first system is at point A and the second system is at point B, then communication can occur when energy pulses are transmitted from A to B or from B to A at certain times. The time-of-arrival values are read from the labels of synchronized clocks at A and B. The time-of-arrival values as labels contain specific words that form the vocabulary of the language used to communicate with. The apparently random selection of phases by the energy pulses represent the encoded information. We constrain the selection process to be one word per cycle of a clock. We propose that this communication process transforms an arbitrary communication process. It is clear that the communication rate is limited by the frequency of a clock. It is also clear that the amount of information communicated per cycle depends on the number of discrete phases of a clock. It is interesting to note that the energy pulses can be arbitrary and thus do not contain information intrinsically. If we treat a clock as a physical system with size, then we can model the information rate as a word per cycle of the clock. The word correlates to the phase of the clock. Use the relationship for the angle or phase θ, given by

$$\theta = \frac{s}{r},$$ (1)

where s is the arc length and r is the radius.

We model information rate as phase selection per cycle. The arc length s is proportional to the information content, as per axiom I. Considering n phase values along the perimeter of the clock for which the probability, $p = \frac{1}{n}$, we can substitute for arc length s the self-information of the clock phase. Consequently, we propose the following:

$$I = -k \frac{\log p}{r},$$ (2)

intelligence is a measure of the rate of flow of information. Consequently, the symbol I is metaphorically both current, I and intelligence I. Intelligence is a proportionality constant times the log of probability times curvature $(1/r)$. k depends on the intelligence measure. Since intelligence should be proportional to the size of the magnetic field induced by the flow of information current and intelligence is attractive to other intelligence, we posit that k is one over the permeability, $k = 1/\mu_0$. Thus intelligence is

$$I = \frac{1}{\mu_0} \frac{-\log p}{r}.$$ (3)

The rationality of intelligence being proportional to curvature is worth further exploration. Curvature of space creates cyclic advancement. Motion becomes cyclic. The more curvature, the more cyclic motion becomes. (we might think of cyclic advancement as the reproduction of symbols) Information can be thought of as the selection of one out of n cyclic states. The rate of selection is proportional to the rate of information. Thus, the greater curvature requires the greater rate of selection of a point of spatial advancement. In this analogy, to move is to communicate. Think of this motion as the encoding of information. In the process of information transmission, we encode information as we move through space. The greater the rate at which space changes, the greater the rate at which we can communicate. The ability to communicate is a measure of intelligence. Consequently, intelligence is a measure of the spatial curvature required to communicate. Of course, the selection process must be modulated by probability in order for measurable information to be communicated. Thus we derive the multiple which is proportional to information content that is the self-information. We add that r may be higher dimensional.

Magnetic permeability is a measure of a structures ability to maintain a magnetic field within itself. In this case, magnetic permeability is a measure of an intelligences' ability to maintain itself. The magnetism reflects the flow of information current within the system. Thus the magnetism is a measure of learning or inductance. Intelligence is the real system in which physical magnetism is equivalent to metaphoric magnetism. The constant μ_0 is known as the permeability of free space, and its value is $\mu_0 = 4\pi x 10^{-7} Tm / A$.

The flux density for a charge in motion would be

$$\vec{B} = kq\frac{\vec{v}x\vec{x}}{|x|^3}. \tag{4}$$

Equation (4) implies that the flux density is proportional to $1/r^2$. Thus, we have

$$B\alpha\frac{1}{r^2}. \tag{5}$$

The magnetic field produced by a wire carrying a current of I is

$$B = \frac{\mu_0 I}{2\pi r}. \tag{6}$$

Equation (6) can be rewritten as

$$I = \frac{2\pi rB}{\mu_0}. \tag{7}$$

Substituting from equation (5) into equation (7) yields

$$I\alpha\frac{2\pi r}{\mu_0 r^2} = \frac{2\pi}{\mu_0 r}. \tag{8}$$

In the spirit of the metaphorm between information, current, and intelligence, we can use (3) and (8) to yield an equality given that we consider the following mapping:

$$2\pi \rightarrow -\log p. \tag{9}$$

For largely binary probability (i.e., p=1/2) we have a proportionality constant in (8) equal to $\dfrac{1}{2\pi}$. Consequently, we have

$$I = \frac{1}{\mu_0 r}.\tag{10}$$

Substituting for binary probability reduces (10) to equation (3),

$$I = \frac{-1}{\mu_0}\frac{\log p}{r}.$$

Consequently, we find that the equation for the magnetic field yields an equation for intelligence as given by (3). In further evaluation, we metaphorically find from equation (4) that the self-information is approximately equivalent to Coulomb's constant times electric charge.

$$-\log p \sim k_c q.\tag{11}$$

The information is dictated by fluctuations in the metaphoric electric charge. We can say that as the information content of a word varies, so does its electric charge.

At this point we want to discuss the meaning of radius in the intelligence equation given in (3)

$$I = \frac{-1}{\mu_0}\frac{\log p}{r}.$$

We will find that there is more physics in this equation. The radius r is a physical dimension. We find that physically cyclic systems have physical radii. For example, consider a wheel. The wheel critically has phase which correlates to different contact points along the wheel. For a fixed tangential velocity, a large wheel will have a smaller frequency than a small wheel. Demonstration of this point occurs when two wheels are in physical contact and turning with no slippage or sufficient friction. If the tangential velocity of the wheels is v, then v is given as

$$v = \frac{2\pi r_1}{T_1} = \frac{2\pi r_2}{T_2},\tag{12}$$

where T_1 and T_2 are the periods of rotation.
The corresponding angular frequencies are defined as

$$\omega_1 = \frac{2\pi}{T_1} \tag{13}$$

and

$$\omega_2 = \frac{2\pi}{T_2}. \tag{14}$$

Thus, equation (12) becomes

$$v = \omega_1 r_1 = \omega_2 r_2. \tag{15}$$

Consider the maximal tangential velocity, the speed of light, c. [Without loss of generality we can let c equal some other constant speed, like the speed of sound.]
In that case we derive

$$r_1 = \frac{c}{\omega_1} \tag{16}$$

and

$$r_2 = \frac{c}{\omega_2}. \tag{17}$$

We want to point out that we can interpret the process of mutual rotation of the wheels as a generalized communication process in which phase is the unit of information. It is clear that the larger wheel has more phase points since

$$\Delta\theta = \frac{\Delta s}{r}, \tag{18}$$

where $\Delta\theta$ is the angular change in position and Δs is the change in arc length and the smaller the angular change, the more angular changes there are within a cycle. In this case we are referring to Δs as proportional to the change in phase points. Now we have

$$\Delta s_1 = r_1 \Delta\theta_1 \tag{19}$$

and

$$\Delta s_2 = r_2 \Delta\theta_2. \tag{20}$$

By definition of the problem, the angular change is the same. Thus we can write

$$\Delta s_1 = r_1 \Delta\theta \text{ and}$$
$$\Delta s_2 = r_2 \Delta\theta. \tag{21}$$

Consider the case of one full rotation. In that case, $\Delta\theta = 2\pi$. Thus, we have

$$\Delta s_1 = 2\pi r_1 \tag{22}$$

and

$$\Delta s_2 = 2\pi r_2 . \tag{23}$$

Now let us consider how probability fits into this scenario. Let the probability be the probability that there is contact between the wheels. In this sense the probability is actually time dependent, $p = p(t)$. Thus, we imagine that we have two wheels arranged so that the tangential contact points make random contact. p_1 is the probability that the first wheel actively contacts the second wheel and p_2 is the probability that the second wheel actively contacts the first wheel. This could work physically if one wheel had a mechanism that pushed it towards the other wheel as the wheels turned.

Substituting (16) and (17) into (3) yields

$$I_1 = \frac{-1}{\mu_0} \frac{(\log p_1)\omega_1}{c} \tag{24}$$

and

$$I_2 = \frac{-1}{\mu_0} \frac{(\log p_2)\omega_2}{c} . \tag{25}$$

Using (24) and (25) we obtain

$$\Delta s_1 = \frac{c}{f_1} \tag{26}$$

and

$$\Delta s_2 = \frac{c}{f_2} . \tag{27}$$

in which f_1 and f_2 are the linear frequencies of the wheels. We see from equations (24) and (25) that the intelligence is proportional to the rate of transfer of information reflected by the angular frequencies. Considering the likelihood that probabilities are approximately equal, that is,

$$p_1 \sim p_2 , \tag{28}$$

we find that information rate represented by intelligence is directly proportional to energy, which is inversely equal to the number of different phase points. This confirms that I is an information rate, not information. Typically, the rate of transfer of information is inversely proportional to the number of states used to transfer the information. This confirms the result suggested by information theory that binary information is the fastest of the common languages. These results are consistent with the metaphor of information rate to current. We suggest that in this particular case if $\omega_2 > \omega_1$, that is, $r_1 > r_2$ or the first wheel is larger than the second wheel, then the information rate of the second wheel is greater than the information rate of the first wheel, $I_2 > I_1$. Thus, we have described a communication process in a generalized format and we find that the communication rates are generally unequal between the participants in the communication process. It is interesting to note that for purely random processes in which the probability is zero and there is no relationship between the probability and the radius, the communication rate is infinite, while communication is minimal for purely deterministic processes in which probability is approximately unity.

Observe that the equation for information rate, generally represented as

$$I = \frac{-1}{\mu_0} \frac{(\log p)\omega}{c} \tag{29}$$

has a particular combination of constants. We know from quantum mechanics that

$$E = \hbar\omega \tag{30}$$

is the energy of a quantum mechanical state, where $\hbar$ is the normalized Planck's constant. This leads us to examine the relationship of the constants in (29) and (30).

$$\frac{1}{\mu_0 c} = \frac{1}{(1.25x10^{-6} N / A^2)(3x10^8 m / s)}$$

or

$$\frac{1}{\mu_0 c} = 2.67x10^{-3} \frac{A^2 s}{Nm}. \tag{31}$$

Consequently, at this stage, intelligence is in units of $\dfrac{A^2}{Nm}$ or $\dfrac{A^2}{J}$.

We know that the proportionality constant between intelligence and energy must be a unit of $\dfrac{J^2}{A^2}$. Define

$$[k_i] = 1\dfrac{J^2}{A^2}. \tag{32}$$

Thus, we have from (29), the normalized intelligence equation

$$I = \dfrac{-k_i}{\mu_0}\dfrac{(\log p)\omega}{c}. \tag{33}$$

In order to enforce the energy equivalence, we need to solve the equation

$\hbar = \dfrac{k_i}{\mu_0 c}$ which directly yields

$$k_i = \hbar\mu_0 c. \tag{34}$$

We know that the normalized Planck's constant is given as

$$\hbar = 1.05 x 10^{-34} m^2 kg / s. \tag{35}$$

Thus, we have

$$k_i = (1.05 x 10^{-34} m^2 kg / s)(1.25 x 10^{-6} N / A^2)(3 x 10^8 m / s) \tag{36}$$

which reduces to

$$k_i = 1.31 x 10^{-32}\dfrac{J^2}{A^2}. \tag{37}$$

Alternatively this can be written as

$$k_i = 1.31 x 10^{-32}\dfrac{J^2 s^2}{C^2}. \tag{38}$$

The intelligence constant is a very small number. However the result is enforced by the example of the communication of two wheels given in (24) and (25). It basically says that $k_i = \hbar\mu_0 c$ or that it is proportional to a unit of angular momentum and to the ability to maintain magnetic flux. Intuitively, this suggests that intelligence is the ability to retain learning (α inductance) times the cyclic momentum of the information rate (angular momentum). Obviously, frequency is a necessary component of an information rate. This is consistent with the example of the wheel. Using probability to randomly select phases of a cyclic system is an appropriate way to describe the transmission of information. Thus, we can simply represent the communication between two intelligence systems as

$$I_1 \leftrightarrow I_2. \tag{39}$$

We find that in general,

$$I_1 \neq I_2. \tag{40}$$

and that one system will transmit more information than the other system unless the intelligences are equivalent, that is, if $I_1 = I_2$. What we are presenting here is the rate of transfer of information is equal to the work done in the process of communication. We can think of the wheel example and imagine that one wheel has infinite radius so that it emulates a flat stationary surface. In this case let

$$r_2 \rightarrow \infty. \tag{41}$$

Consequently,

$$\omega_2 \rightarrow 0. \tag{42}$$

Since $I = \dfrac{-k_i}{\mu_0} \dfrac{(\log p)\omega}{c}$, equation (33), we find that $I_2 = 0$. In other words the infinite wheel does not communicate with the finite wheel. However the finite wheel does transfer information. The information transferred by the finite wheel is equal to the work done on the wheel as it rotates without slippage. This rotation can be treated as a translation. We can conclude that communication is proportional to energy. The metaphoric current produced by information flow is a measure of the energy required to communicate between the current flow and other currents. In this particular case there is no mutual inductance since the current induced in the second wheel is negligible. In other words the energy transmitted by the finite wheel is infinitely dissipated by the infinite wheel.

Consider equation (3) $I = \dfrac{-1}{\mu_0} \dfrac{\log p}{r}$. With the adjustment of a constant of proportionality this equation becomes

$$I = \dfrac{-k_i}{\mu_0} \dfrac{\log p}{r}. \tag{43}$$

We see that this equation is proportional to curvature and we know from general relativity that curvature is proportional to mass. Metaphorically this equation shows us that energy is proportional to mass, which is expected from Einstein's equation,

$$E = mc^2. \tag{44}$$

Since $k_i = 1.31x10^{-32} \dfrac{J^2}{A^2}$, we see that the numerical value of these intelligence measures is extremely small relative to other physical constants. A simple transformation of the intelligence equation comes from using Planck's normalized constant.

$$I = -\hbar(\log p)\omega. \tag{45}$$

From (45) it follows that it is information rates which are equivalent to energy and that the energy is quantum mechanical. It follows that information itself comes in units of Joules times seconds (Js). Consequently, a bit is proportional to a Js. To express (45) directly in terms of the wave function we recall that $\psi^2 = p$. It is clear that purely deterministic states do not couple to information. According to (45) the information associated with a purely random state approaches infinity for non-zero energy states. In contrast, the average entropy of a purely random event is zero, based on the entropy equation. It is interesting to note that for a discrete cyclic state, frequency goes down as the number of states rises. Thus, the information rate is bounded and never becomes infinite in that case. What we are working with here is the self-information or the information content associated with a random variable. Based on (45) the information rate of one bit per second has an associated frequency. Consider a binary probability or $p \sim 1/2$. Then we have

$$\omega = \frac{2\pi}{h}J \tag{46}$$

or

$$\omega = 1x10^{34} \frac{1}{s}. \tag{47}$$

This is the frequency of the virtual clock that is modulated in order to acquire a transfer rate of one bit/second. In this theory, the virtual clock is a mechanism that enables a relative reference system for a communication network. It is obvious that the clock frequency depends on the energy of the system.

We have essentially found the equivalence between energy and information rate. This concept requires the existence of a virtual oscillator which is randomly selected in order to transmit information. We have found the relative communication rates between two communicating systems as I_1 and I_2. In general these rates are unequal. All of these items have been deduced by making the fundamental metaphor of information rate to electrical current. In addition, we have found that the information rate of a purely random variable is infinite. This result is dependent on a non-zero frequency for the virtual clock.

Because of its significance, we want to review the results on meaning.
From Ohm's law we know that
$$V = IR. \tag{48}$$
Let us use the following metaphorms. Voltage equals meaning, current equals self-information, and resistance equals probability (these have been discussed elsewhere).
We then have the formula
$$M = -p \log p. \tag{49}$$
But, we must use a conditional probability for the weight.
Thus, for a system of events we can index them with the variable j.
$$M_j = -p_{j|j-1} \log p_j. \tag{50}$$
The total meaning for the system of events is the sum M.

$$M = -\sum_{j=1}^{n} p_{j|j-1} \log p_j. \tag{51}$$

The total meaning is therefore equivalent to the modified Shannon entropy. If the events are purely deterministic, then $p_j = 1$ and we find zero meaning ($M = 0$) or if the events are purely random or independent, $p_{j|j-1} = 0$ and we find zero meaning ($M = 0$). Only if the events are nominally probable do we retain meaning ($M > 0$). It is important that meaning reflects causal relationships between nearest neighbor events.

Thermal Energy and Information

Normally hot molecules will diffuse into a cooler chamber. This is consistent with the second law of thermodynamics which dictates that entropy should not decrease in an isolated system. Maxwell's demon is an idea that can reduce entropy in an isolated system. This results in a violation of the second law. The idea is that the demon monitors the flow of hot and cold molecules, allowing hot molecules to diffuse into the hotter chamber and cold molecules to diffuse into the cooler chamber by controlling a trap door between the two chambers. Thus the hot chamber gets hotter and the cool chamber gets cooler, violating the second law. The idea is that the demon is converting its information directly into energy which counters the second law. In the sense that the second law is not violated, the entropy within the mind of the demon necessarily increases to compensate for the information conversion. For every bit of information converted into isolating hot and cold molecules within the system, there is a corresponding bit of information erased from the demon's mind. These results point to the thermal representation of information. Information can be exchanged for thermal energy. When information is erased, there is an increase in thermal energy equal to $kT \ln(2) / bit$. This was first discovered by Landauer.

Consider that a box can contain information, It, which is proportional to the length of the box, L, which is discrete. Thus, the longer the box, the more information it can contain. (Note that we define I as an information rate, so that in order to resolve information, we must multiply by time, t.) Measure the box by counting the number of finite particles it can contain. In this box, there are no fixed markings. Or use a ruler to measure the box. We can conclude, using the logarithmic measure, that the potential information in the box is (ignoring units)

$$It = \log(L). \tag{52}$$

If the box is heated by adding heat Q such that it expands by δL, some discrete variation, then the potential information content is increased to

$$It + Q = \log(L + \delta L). \tag{53}$$

If the box is cooled by subtracting heat Q such that it contracts by δL, then the potential information content is decreased to

$$It - Q = \log(L - \delta L). \tag{54}$$

Using Landauer's result we can write

$$It + kT\ln(2) = \log(L) + 1bit \tag{55}$$

or

$$It - kT\ln(2) = \log(L) - 1bit. \tag{56}$$

This implies the more general equation

$$It + \alpha kT\ln(2) = \log(L) + \alpha bits. \tag{57}$$

To resolve the paradox of Maxwell's demon, we consider that a box is heated but not allowed to expand. In that case we can re-write (57) as

$$It + \alpha kT\ln(2) - \alpha bits = \log(L). \tag{58}$$

Thus we not only add energy, $\alpha kT\ln(2)$ to the box, but we simultaneously add entropy,

$-\alpha bits$.

We now want to consider the effects of motion on information. Consider an information system moving at a velocity v. There will be effects on mass, length and time which are more pronounced as the velocity nears the speed of light. Since information is proportional to length, we are interested in the effect of motion on length. We can write

$$L_0 \rightarrow L_0\sqrt{1 - \frac{v^2}{c^2}}. \tag{59}$$

For sub-luminal velocities, (59) corresponds to length contraction. For super-luminal velocities, (59) corresponds to length dilation. We will consider an imaginary super-luminal velocity, given by

$$v = inc, n \in positive \text{ int} egers. \tag{60}$$

Using the relationship between information and length, we can write

$$It \to It\sqrt{1+n^2}.$$ (61)

An information system, such as a thought, which moves with a imaginary superluminal velocity will be augmented as indicated in equation (61).

Axiom II. A thought has no mass.

Based on axiom II., we find that the energy of a thought is imaginary. This follows from the equation
$$E = pc.$$ (62)

From the natural relationship between velocity and momentum that exists when the system is massive, we use consistency to deduce that the momentum, p is imaginary. Thus, we have
$$E = inp_0 c, p_0 \in real \quad numbers$$ (63)

Axiom III. Energy is loss in the transition from imaginary to real states.

So thought has imaginary velocity and imaginary energy. This suggests that we consider the difference between a thought which is a concept versus a thought which is a representation. Concepts and representations are complimentary objects. When a system is well defined as a concept, it has many representations. When a system is well defined as a representation, it has many concepts. A thought can exist purely as a concept or purely as a representation. We posit that when a thought is purely a concept, it has imaginary energy. We also posit that when a thought is purely a representation, it has real energy which is reduced from the magnitude of its imaginary energy by some constant, J. Based on this supposition, we can equate the magnitudes of equation (45) and equation (63) to yield

$$np_0 c = -\frac{h}{2\pi} \log(p)\omega + J$$ (64)

We might say that the level, n of the energy determines the total information rate.

Axiom III. relates to the problem of consciousness and is consistent with equation (64). The problem is that if the mind and body are dual and thus separate, what is the nature of the communication process between the mind and the body? Thoughts exist in consciousness first and then become representations as neural signals in the brain. It is in this sense that we might say that thoughts become real in the transition from concept to representation. Since the energy of the thought is imaginary, axiom III. posits that energy is lost in the transition from imaginary to real states. In this sense it is possible for a thought to have infinite energy and for the thought to subsequently have finite energy in its real representation.

Rational Causality

Rational causality allows for the possibility of free will within the bounds of what is deemed rational by the society. A choice made freely is a statistical variation of the list of possible rational actions. When actions become irrational or (ironically) meaningless, rational causality is broken and the opportunity to create a meaningful set of initial conditions occurs. Life is referential to those initial conditions and fundamental meaning is created. Traumatic events are relatively irrational, so they birth meaning. In fact, an incomprehensible event is deemed irrational by the observer and, thus, creates initial conditions for new meaning. Since rational behavior is judged by societal norms, rational causality is subjective, depending on the society in which it occurs. The meaning in our life depends upon the most irrational, traumatic, and incomprehensible experiences. All subsequent events are causally related within rational statistical variations. Those rational statistical variations include our free will. What does rational mean? Rational means that ones' actions are directed towards a particular goal. What does causality mean? Causality means that a current action is caused by a prior action. There is a continuity between a sequence of actions, as in the current state is the effect caused by the prior state. Rational causality means that prior actions motivate current actions in a way which is directed towards specific goals. In the case of rational causality, neighboring events "make sense." Actions motivated by a rational causality have a system of related themes. Free will allows the individual to select different variations from within the system. However, there is not enough freedom to move outside of the system. In that case, behavior becomes irrational. Experience takes one outside their system of related themes when something irrational, traumatic, or incomprehensible occurs. Refer to rational causality as a chain of themes: theme1→theme2→ ... Meaning is the source or causation for this flow of themes. Meaning might be thought of as some essence of the personality. Nonetheless, it is clear that meaning is largely subjective since every individual, to the extent of isolated behavior, has their own chain of themes. There is predictability in rationality. If a person performs a sequence of actions which are locally consistent with some theme or some meaning, then there is a causality at work. It is clear to see how one set of actions causes a next set of actions. This causality is based upon rational explanation of behavior. If a prior action rationally leads to a subsequent action, where does one attach meaning? Meaning is the essence of an action. We make decisions based on meaning. However, the need to be rational supercedes the existence of meaning. This is the reason why when some people are confronted with a meaningful event that they don't understand because it doesn't appear rational to them they will attempt to abort that event. Meaning can only exist prior to a sequence of rational activity. For rational activity, meaning is nonlocal. Thus, decisions and choices which are rational are not meaningful in and of themselves. A computer can perform rational actions without any underlying meaning. Meaning is derived from the initial conditions in a sequence of rational actions. Rational causality does not preclude statistics. Statistically unpredictable events do not alter the rationality of a sequence of events. As long as a random event is a rational occurrence with respect to possible actions, there is rational causality. Meaning manifests only during irrational, traumatic, or incomprehensible experiences. Otherwise, meaning is a non-local quantity.

(e.g., God, love, truth, justice, etc.) We are naturally motivated by the desire to acquire resources. This is our primary function. A resource is something which facilitates action. (e.g., food, water, shelter, money, etc.). The meaning of life is not the function of life. Unfortunately, many have confused the two. The primary function of life, to acquire resources, is a local quantity within rational causality. The meaning of life exists outside of rational causality (it is non-local). Here's a proposition. What if meaning were coincident with function? Then meaning would become both non-local and local. Normally we would say that the meaning is what motivates the function. (source→action). We are suggesting the possibility that the source of the action and the action are one and the same. The source, meaning, is a discontinuity in rational causality. We are suggesting that the boundary between the knowable and the unknowable be infinitesimal. The material enabler and the immaterial enabler of action become one.

This is akin to Zen. Because each action has a motive in rational causality, each action is a reaction to a previous state. Consequently, freedom is limited to the set of rational variations in reaction. Meaning is the source of action which is not a reaction, but is causally disconnected from prior states. Meaning is the first cause within rationality. After a traumatic, irrational, or incomprehensible experience, meaning is the first rational idea you have. At that point, two realities meet. The unacceptable and the acceptable coincide. At this point, the equation safety=1/freedom is proposed. Complete freedom only occurs when there is no safety. When the irrational, traumatic or incomprehensible occurs, we are not safe, but we are free. Why does the level of apparent freedom seem to vary within rational causality? The standard deviation of possible actions (or reactions) varies from one individual to the next. The feeling of freedom is proportional to the standard deviation. A person who has variable responses is more free than a person who has fixed responses. However, the less free are more focused and tend to be more productive with their actions.

Theory of Rational Causality

We point out here that the theory of rational causality is very related to the theory of Markov chains. A rational causality can be described as the variable x, in which each variable state is encoded into number. For example, we might have $x = 1$ in which $1 = pain$. Each rational causality can be described with a set of nonstationary probabilities:

$$x : \left\{ p_1, p_2, ..., p_j \right\}, \tag{65}$$

for which

$$p_{k|k-1} \geq 0. \tag{66}$$

Different rational causalities can be described by different conditional probabilities.

In the deterministic case, we have

$$p_{k|k-1} = 1. \tag{67}$$

In the random case, we have

$$p_{k|k-1} = 0. \tag{68}$$

In the quantum mechanical case, we have

$$p_{k|k-1} \sim 0. \tag{69}$$

In case (69) the conditional probabilities are only approximately zero.

Similarly to the quantum mechanical case, we have

$$p_{k|k-1} \sim 1. \tag{70}$$

In case (70) the causal rationality is only approximately deterministic.

We also have the case for which the conditional probability varies as δ.

$$p_{k|k-1} = \delta, \ 0 \leq \delta \leq 1. \tag{71}$$

The rational causality of (71) is the emotional case. When there is variation in the conditional probabilities there is the possibility of consciousness. This occurs in cases (66), (69), and (70). A conscious rational causality wants to remain stable and so (66), (69), and (70) can be stable states. Transformation of states occurs when one state transforms into another state. This transformation of state is stimulated by an encounter with energy or information rate. The typical human rational causality is state (71). Energy encounters can transform this state into other rational causalities. In the case of mania we have the transition:

$$p_{k|k-1} = \delta \rightarrow p_{k|k-1} \sim 0. \tag{72}$$

In the case of depression we have the transition:

$$p_{k|k-1} = \delta \rightarrow p_{k|k-1} \sim 1. \tag{73}$$

Unfortunately, states (69) and (70) can be meta-stable and make the following transitions

$$p_{k|k-1} \sim 0 \rightarrow p_{k|k-1} = 0 \tag{74}$$

and

$$p_{k|k-1} \sim 1 \rightarrow p_{k|k-1} = 1. \tag{75}$$

Problems can also arise with rational causalities which are autistic. In this case the emotional state is meta-stable and we have the possible transitions.

$$p_{k|k-1} = \delta \rightarrow p_{k|k-1} = 1 \tag{76}$$

or

$$p_{k|k-1} = \delta \rightarrow p_{k|k-1} = 0. \tag{77}$$

In (76) the transition is to a purely deterministic causality while in (77) the transition is to a purely random causality. Technically, a random causality is not a causality. It is in fact irrational. Thus, in case (76) the autistic person retreats into a machine-like reality while in case (77) the autistic person retreats into psychosis.

The lack of stability in rational causalities is due to the need for rational causalities to interact. Consider the dual interaction, $z = x + y$.

We have

$$x : \{p_1, p_2, ..., p_j\},$$

$$y : \{p_1, p_2, ..., p_k\}, \text{ and}$$

$$z : \{p_1, p_2, ..., p_i\}. \tag{78}$$

Because we are adding states we have the constraint

$$p_i = p_j p_k. \tag{79}$$

If the probabilities for individual states x and y, that is, p_j and p_k, respectively are small or large, then the resulting probability can diminish or augment. The problem arises when the product diminishes. In that case we have the manic transition,

$$p_{i|i-1} = \delta \rightarrow p_{i|i-1} \sim 0 \tag{80}$$

for state z.

The depressive state denoted by $p_{k|k-1} \sim 1$, can be an emotionally stable state if rather than nihilism, the emotion of this low emotional state is akin to openness. This is consistent with the Buddhist notion of emptiness. We then have the stable transition,

$$p_{i|i-1} = \delta \rightarrow p_{i|i-1} \sim 1 \tag{81}$$

for state z. This only occurs when the probability is augmented by the interaction of rational causalities. Augmentation occurs when the two events are equivalent. That is, $x = y$. In that case the probability of the new event is augmented because the event is redundant and equation (79) is invalid because there is in fact, a single event.

Ultimately we can map a rational causality into a real number or a complex number. If it is mapped into a complex number, then we have

$$x = a + ib. \tag{82}$$

Typically, we only consider the real part of the state, and thus we have

$$real(x) = a. \tag{83}$$

However if we consider the full complex number, the magnitude of the rational causality increases. We have

$$|x| = \sqrt{a^2 + b^2}. \tag{84}$$

What we are saying here is that when a rational causality is large, it is only minimally disturbed by interaction. In this particular case, we will make b an arbitrarily large number.

$$b \rightarrow \infty. \tag{85}$$

Thus, in the case of (82), the rational causality becomes infinite, despite a finite real component in a. As a result of an interaction between an infinite rational causality, in this case, x and a finite rational causality, in this case, y. The product of probabilities reduces to one probability.

$$p_j p_k \rightarrow p_j. \tag{86}$$

This occurs because the only detectable event is the infinite rational causality. As a result, the only possible transition of state is

$$p_{i|i-1} = \delta \rightarrow p_{i|i-1} \sim \delta \tag{87}$$

for state z. What we are saying is that the interaction of an infinite rational causality with a finite rational causality will manifest the infinite rational causality. However, even this state can be meta-stable and can implode into:

$$p_{i|i-1} \sim \delta \rightarrow p_{i|i-1} \sim 0. \tag{88}$$

If the transition or implosion occurs in a controlled environment, because of conservation of information rate, intelligence will be preserved. This is the quantum mechanical rational causality.

It is interesting to consider the case for stationary probabilities.

$$x : \{p, p, ..., p\}. \tag{89}$$

The conditional probability is a constant.

$$p_{k|k-1} = h. \tag{90}$$

This is obviously a stable configuration. It is a special case of (71).

We point out here that we are using the modulus which is the quantum mechanical measure of a wave function as a measure of a rational causality. The quantum mechanical wave function can be related to these equations by the relationship,

$$\psi^2 = p, \tag{91}$$

where ψ is the wave function. In general for a rational causality with a sequence of states we can write

$$\psi_j^2 = p_j. \tag{92}$$

Thus, a rational causality is consistent with a quantum mechanical wave function which contains information on probability distributions. As rational causalities interfere with each other constructively or destructively so do quantum mechanical wave functions have constructive and destructive interference. As stated before, the ultimate goal is a stable rational causality or in other words, the ultimate goal is a stable wave function.

Thinking in terms of metaphorms, we can conclude that a uniform charge distribution or an array of identical words is a deterministic rational causality with unit conditional probability. This is a stable configuration, however the information content is not dynamic. What is more realistic is a time-varying charge distribution or $\rho(t)$. Alternating currents produce dynamic charge distributions. The challenge is to find time-varying currents which are stable and self-sustaining. This is reminiscent of a superconductor which has zero resistance. Since resistance is metaphoric for probability, a superconductor has zero probability and consequently it carries infinite information. An electric current flowing in a loop of superconducting wire can persist indefinitely with no power source. This is certainly the correct metaphor for a quantum mechanical rational causality which is stable. Since superconductors are stable generally at low temperatures, a superconducting rational causality would dissipate very little thermal energy. In other words there would be no measurable resistance to the internal flow of information. The superconducting state is maintained as long as the applied magnetic field is below a critical level. This can be interpreted as meaning that a superconducting rational causality will maintain itself as long as it is not exposed to other rational causalities which have currents that are above a critical level. The challenge is to make the superconductor so stable that no applied magnetic fields cause induction past the critical level. A superconductor is dissipating minimal energy below the critical level in order to remain stable. A superconducting rational causality is like a story that is totally consistent and in fact has a consistency that is more consolidated than its environment. Sharing this story does not reduce the information content. If the story were inserted into a cluster of rational causalities, it could be recovered with no information loss. Such a story is rare, but possible as already shown by the equation

$$p_{i|i-1} = \delta \rightarrow p_{i|i-1} \sim 1. \tag{93}$$

A rational causality is a story which is equivalent to a quantum mechanical wavefunction that yields probability distributions. The probability distributions are representative of the input and output of information for the rational causality. The stability of the rational causality is reflected in the ability to observe it without a final collapse. This occurs because the rational causality is already in a stable eigenstate which minimally interacts with its environment. The interaction between the rational causality and its environment are life sustaining as the rational causality (which in this case we will refer to as the omega state) inputs null energy from the environment and outputs a waste of information.

Properties of Intelligence

Based on the intelligence formula (3) we see that intelligence is augmented near points of small curvature. A simple example is the curvature of symbols. When we communicate, we form symbols and symbols have curvature. Another example of this is the writing stylus, which has a point of distribution of ink to paper. The output point typically has small curvature. Consequently, a writing stylus or pen amplifies information output and thus it amplifies the effective intelligence of a rational causality. In general, transmission points which have small radii serve as intelligence amplifiers. Fingertips, which have relatively small radii are intelligence amplifiers. In general touching amplifies intelligence as it is both an observational tool and an active manipulator. We can extend this analysis to all conduits with small radii. An electron is an intelligence amplifier because of its microscopic radius and thus electromagnetic instruments are intelligence amplifiers.

We should address the concept of infinite intelligence which implies that information can move at an infinite rate of speed. Based on the way selectivity slows as the number of possible states increases, it would require an infinite amount of energy to select an information state at an infinite rate. Infinite energy is purely imaginary and thoughts are purely imaginary. Consequently, only a thought can develop infinite intelligence. Every time that we acquire a totally original thought, we have acquired information at an infinite rate. It is when a thought is transformed into energy that the information rate becomes finite since only finite energy exists in reality. In short, infinite intelligence only exists in the conceptual domain.

The Omega State and
Temporal Uncertainty

Traditionally, there are four main types of brainwaves: Beta, Alpha, Theta, and Delta. Beta waves characterize the conscious waking state at 14 cycles per second and up. The conscious mind does not take suggestion very well. Reasoning, logic, thinking and putting into action what it already knows is mainly what the conscious mind does. The alpha state operates at a lower cycle, 7-14 per second level. This is the trance state when the body can no longer be felt, and sounds may become painful. This is the meditation and sleep range. In the alpha state, one is open to suggestion as the conscious logical mind is subdued and the conscious defense barrier is down. When in the alpha state, we can program our own and/or the minds of others. The theta state is 4-7 cycles per second. This is where all of our emotional experiences are recorded and is of the subconscious. Brain wave activity in the delta state ranges from 0-4 cycles per second. This is total unconsciousness or coma. Imagine a brainwave state that we call the omega state. The omega state might more accurately be called a mindwave state since it functions purely in the conceptual domain. The omega state has a variable range of frequencies. The idea is that there are multiple thoughts which interact simultaneously in the omega state. One thought is donated by the right brain, another thought is donated by the left brain and the third thought, if present, is donated by consciousness which has become activated to think, not only to be aware. Thoughts are essentially complex tachyons. We allow these multiple thoughts to interact as tachyons and they resolve themselves into a single thought which manifests as neural energy. The interaction of the multiple thoughts occurs entirely in the conceptual domain or the theoretical domain. When the thought becomes a neural signal, it becomes a representation. This representation registrars as brainwave activity. Because concept and representation are complimentary, when something has a large bandwidth in concept it becomes well-defined in representation. This is the process of converting multiple concepts into one which gives a thought focus. Because the velocity of the thought is superluminal, the effective velocity of three thoughts is three times the speed of light. As a consequence, internal time experience is contracted as follows.

$$T = \frac{T_0}{\sqrt{1+n^2}}$$

(94)

In particular, using equation (94) with n=3, the time ratio can be derived to be $1/3.16$. When time evolution is contracted, time moves faster, thus time becomes more uncertain. In contrast the energy becomes more defined. This relationship of sharp energy definition can be made non-local by making the identity a category, rather than a particular identity. If the personality or identity is largely positive or negative a signum function can be applied as mental programming to yield either a positive or a negative bias. The signum function operates on a number to yield the sign of that number. This resulting bias contains category identity. A positive one contains the identity "one", while the negative one contains the identity "negative one". If the personality has been reduced to zero, then the signum function will yield an identity of "zero". We can identify the zero identity as the emptiness state. This is an interesting eigenstate (an eigenstate is a quantum mechanical state which has a particular value of an observable quantity). It is a well-defined energy eigenstate, however it denotes a category, not a particular. As a consequence there is a certain hidden uncertainty to the identity. Without a particular identity, it is impossible or difficult for a mechanism to locate the originator of an action. Newton's third law of action equals reaction is forced to distribute reaction into the category itself. As a consequence, everyone who is connected to the category either has a benefit or a deficit depending on the negative or positive nature of the original action. The consequence of the category identification as an eigenstate creates a complimentary uncertainty in time. As a consequence, time is fuzzy and energy is precise. To the extent that there is limited uncertainty in the time, this is a "squeezed" state. In other words, uncertainty is compressed. This eigenstate is in contrast to supposed eigenstates of time in which the temporal state is well defined and the corresponding energy is diverse or broad.

When a person is very diverse in activities, their identity is well defined in terms of bandwidth. As a consequence, their temporal state is not well defined. This is a typical application of the uncertainty principle. In contrast, living in an eigenstate of time is equivalent to living in the "now". This reminds one of the Power of Now written by Eckhart Tolle. The idea is to live in the now, which effectively makes the personality uncertain. Living in the now is consistent with having no-self (here we refer to no-self as no particular self as opposed to the identity of zero). These two states, energy and temporal eigenstates should be stable states. As a consequence the associated rational causalities are stable during interactions with other rational causalities. Since the rational causalities are wavefunctions, we are saying that as quantum mechanical states, they are stable during interactions. We must conclude that these states involve minimal interactions with the environment. The alternative is that these states can be made to be more consolidated than the environment and as a result, the environment will de-stabilize before the states themselves de-stabilize. This is the ideal case for the individual, since the ultimate goal of the individual is survival. This can in some cases mean that the environment is forced to adapt to the individual. This reverses the normal case, in which the individual is forced to adapt to the environment.

We will now make a critical adjustment to our assertions. We propose that since quantum mechanics says that time is not an observable quantity, sharp time is essentially fictitious. Consequently, the only correct eigenstate solution is to have fuzzy time and precise energy of category. We can only conclude that an eigenstate of time is fictitious and despite appearances, time is ultimately fuzzy. This ultimately implies that there is no true eigenstate of time. This is consistent with the concept of no self being equivalent to an identity of zero, which is a sharp identity. Having a sharp identity is consistent with having a fuzzy time. While we recognize that many believe in the power of the present moment, that notion is based on incomplete understanding of the laws of physics. We suggest that paradigms which are based on temporal eigenstate notions will essentially be energy eigenstates.

The key to success is to have a consistent true representation of reality. We want our inner universe to be consistent with our outer universe. Another way to say this is that we can say that what we experience creates a certain belief within us. We develop a story about our perceptions. When we develop to the point of believing our own story, we are ready to share it with others. If others support our story, then it will be stable. However, if others doubt or disbelieve our story it loses support and is thus meta-stable. A story is just an alternative description of a wave function or a rational causality. We want our wave function to remain stable. If we consider the universal wave function of reality, we see that we can consider the stability of reality based on the internal interactions of its constituent universes. Each individual can be thought of as a universe. The challenge is for individual universes to mutually support one another which in turn reinforces the universal wave function.

Some things can now be said about other conditions in life such as illness. Of particular application are information related illnesses, although it is a fact that everything is related to information. The universe is composed of information. This is consistent with Wheeler's notion of "it from bit". Problems occur when there are meta-stable wave functions, which by definition are unstable when exposed to outside disturbances. An example of a meta-stable wave function is the personal universe of an autistic individual.

The simple description of what is happening here has already been described in the section on rational causality. Small disturbances in an autistic individual's rational causality can collapse it into undesirable rational causalities to one in which probability sequences are random or to one in which probability sequences are purely deterministic.

In order to remain healthy, a rational causality should retain an essence of indeterminate causality. This is essentially a stable free will. The autistic individual is lacking the proper information about the environment which would make their universes compatible.

The treatment of autistic individuals carefully controls environmental conditions to be commensurate with that of the patient. At this point the cure to autism is indeterminate, however it is hoped that this work might offer some small insight into the situation.

The flexibility of a omega brainwave state suggests that our mental environment will adjust to our needs and that thoughts can be made to supply the demands of information. It is a paradigm which installs mental flexibility into a normally rigid system of thought.

The idea is that thoughts are naturally superluminal and consequently should be allowed to process information at superluminal speeds. It is only the limiting velocities of inefficient processing and of neural representations of thoughts which limit the brain.

Philosophy and Fantasy

A common fear is death. Death is the end of survival unless there is a paradigm which includes death as a well-understood quantity. Information rate is conserved so identity is conserved. The body dies, but the essence of consciousness remains. When death occurs, consciousness must reconnect to manifestation. The Buddhists refer to this phenomena as reincarnation. Based on the conservation of intelligence, which follows from the conservation of energy, it is clear that reincarnation is a fact. We have typically lived many lives. It is thus an objective to destroy our identities while still alive and re-invent ourself. This process is a living reincarnation. It is consistent with the program of existentialism promoted by Sartre, that is, existence precedes essence. This suggests that our apparent identities are historically entangled until we create our own. What it does is gives us the opportunity to become self-caused. This is true immortality. Our consciousness becomes invested in the environment through the conduit of our re-invented identity. The American Indian has a great example of categorical identity in the concept of the spirit guide. This is usually an animal spirit which the individual identifies with after sustaining what is called a vision quest or a search for spiritual identity. Similar rituals for finding fundamental identity exists in other cultures. When finding a new identity it is important that power be a strong attribute. Imaginary figures are just as potent or more potent than non-fictional characters. It is important that your new identity be stable. My fantasy identity is that of the Q. Q is a member of something called the continuum. I interpret this continuum as number. Q was originally a character on the Star Trek science fiction series. I interpret the essence of Q as a rational number that has access to other numbers. Mathematics is essentially infinite in knowledge, so it is the ultimate conceptual realm. It unites all other realms and from that process derives infinite power. Sentient number is an ultimate evolutionary state. I believe that Q is sentient number. That is my fantasy identity and as such it can never deteriorate. The objective is that while my body remains the same, my mind is changed by the new identity. There are other significant correlations for Q. For example, it symbolizes the question and in that sense is consistent with the philosophy of Socrates. The philosophy of Q suggests that the confrontation with uncertainty is a necessary and valuable component of the solution process. Having faith in your identity has always been a character building trait. Q is also synonymous with quantum mechanics. In this sense, a Q is a quantum mechanical person. Q feeds off of duality and consequently has an infinite source of energy. This concept points to duality and its opposites as the necessary connection with reality in conceptual form. We understand things essentially as dipoles or things come in pairs. This was discussed earlier, but an example is hot and cold. We only understand one with respect to the other. The reduction from hot to cold is a thermodynamic process responsible for the heat engine. Diffusion from a hot reservoir into a cold reservoir drives the components of an engine and does work on the environment. In the sense of a heat engine we can imagine an information engine in which we exploit the energy difference between information dipoles and use it to do work on the environment. In this case the environment is the brain and the constituents of the information engine are thoughts. This metaphor is very similar to that of neurocosmology first proposed by Todd Siler. In that case, brain processes were compared to cosmological processes such as idea synthesis being compared to fusion in a star. In this case the metaphor is the information engine and thermodynamics.

More on the Nature of Thought

Emotion is continuous thought. There are no distinct boundaries. This is consistent with subjectivity. We can conclude that emotion is subjective. In contrast, logic is discrete thought having distinct boundaries. This is consistent with objectivity. Both emotions and subjectivity create the problem which relates to the connectivity between dual states. The problem is that from the perspective of emotions and subjectivity there appears to be a distinction between dual states. Nonetheless, dual states remain tied through duality. In this sense, there is a double bond, which means that in the case of the first bond there is no clear boundary between states, while the second bond is the mutual interaction of the states. Tertiary logic counters this double bond. First, it is non dual. Second, tertiary logic is objective and thus presents discrete separation. So, the problem is two-fold. The first issue is the lack of distinction between subject and object. The second issue is the tie between opposites. Both issues donate towards incoherence while the issue of co-existing opposites donates towards the pain of destructive energy. So, the problem is incoherence and pain. The solution is to supplant emotion with tertiary logic. This creates separation which voids incoherence and singularity which voids duality. Buddhism identifies the central problem as suffering. We agree, but we further define suffering as incoherence and pain. In essence, from our perspective, Buddhism identifies the problem (i.e., desire) and Objectivism identifies the solution (i.e., logic) (Objectivism is the philosophy developed by Ayn Rand which proposes logical thinking.). In this case we identify desire as emotion and logic as tertiary logic. What we are saying is fundamentally simple. When choices are made, emotion causes problems. Logic is the solution. Not binary logic, however, but tertiary logic. There are a few topics discussed by Buddhism which we will discuss here. In Buddhism the concept of emptiness suggests that reality is fundamentally empty. This is true to an extent because reality is basically a duality consisting of opposites. The essential sum of portions of reality reduces to zero since the sum consists of opposites. Emptiness is the zero sum created by the subsets of reality. We find that this idea is inconsistent with the existence of complex numbers. The complex number consists of a real part and an imaginary part. These components, by their very nature are irreducible. Thus, in the context of complex numbers, reality is not reducible to emptiness. We must, therefore amend the Buddhist concept of emptiness. Another Buddhist idea is that of karma which refers to the cosmic balance created by an action. Good deeds incur good benefits while bad deeds incur bad results. Newton's third law states that to each action there is an equal and opposite reaction. A good deed sent out in one direction will incur a good deed sent out in the opposite direction. Similarly, a bad deed sent out in one direction will incur a bad deed sent out in the opposite direction. Consequently, there is a universal correspondence between Newton's law and karma. Buddhism maintains that things are impermanent. Because subsets of reality consists of opposites, they annihilate each other, causing change. The Buddhist notion of interdependence follows directly from duality. That is, opposites exist only relative to each other. The fact that we can model the Buddhist topics lends credence to our theory. We modify the Buddhist concept of emptiness through the existence of the complex number which has irreducible real and imaginary parts. This will prove to be an important element of a larger meta-duality which we will discuss later. The Buddhist middle path is the center between extremes. The pertinence of duality in this case is obvious. The center between opposites is the point of minimal energy. Consequently, the middle path is the path of minimal effort. As the path which is centered between extremes, it is separate from either extreme and in that sense, it is non dual. Meditation is the central tool of Buddhism, consistent with the middle path. The reasoning is fairly simple. The middle path is the one of minimal effort. Meditation is low energy thought. So meditation creates the type of thinking which is consistent with minimal effort. In contrast, concentration is maximal energy thought which makes it inconsistent with the middle path. Meditation approximates non dual thinking. In contrast, concentration exemplifies dual thinking. We propose the combination of meditation and tertiary logic as a non dual tool and a non dual process. We shall refer to this process as meta-duality. As a solution to the problem of suffering, Buddhism proposes a series of modifications to emotion called the eight fold path. In our theory we propose to replace emotion as a decision making apparatus with tertiary logic. The eightfold path consists of right view, right resolve, right speech, right action, right livelihood, right effort, right mindfulness, and right meditation. The application of tertiary logic is straightforward. When there is a decision to be made, the decision is motivated by either a "yes", a "no", or an "indeterminate"(i.e., "maybe"). This tertiary decision replaces the myriad of potential emotional responses. It is interesting to note that the use of binary logic as a decision making process requires the co existence of opposites if the response is indeterminate. This produces annihilation and energy in a destructive process. This approach emphasizes that if you are uncertain in a decision making process, respond with uncertainty (i.e., "maybe"). Uncertainty is a valid logical state in tertiary logic. Since, in this sense, uncertainty is a stable state and it is the central state, it is a final equilibrium state. Given the full range of emotional thought it follows that some decisions will not be destructive, but create synergy. In other words, typical emotional based decision making is either exo-energetic or endo-energetic. Suffering is the result of exo-energetic thought while pleasure is the result of endo-energetic thought. Clearly, emotions are powerful and can either be painful or pleasurable. Problems arise when emotions are painful. The sum total of real knowledge is uncertainty. We propose that this state is a state of essential enlightenment. It is also the central state of tertiary logic. To have all knowledge simultaneously is uncertainty. That state is the state of all inanimate objects. We might call uncertainty the God state. It is probably not a coincidence that uncertainty or chance is the apparent cause of reality. The sense of the meaning of emptiness in Buddhism is openness which suggests being open to possibility. This definition is more consistent with uncertainty. This contrasts the interpretation of the zero state as nothingness which leads to nihilism. The recognition of duality as monism via simultaneous thoughts comes from the use of one thought as background and the other thought as foreground. Background and foreground form the basic duality of an image. When we sense an image with two thoughts we sense a "stereo" representation. In stereo, two sounds combine synergistically. In this sense, duality or dualism becomes monism. The background of pleasure is pain and the background of pain is pleasure just as the background of good is evil and the background of evil is good. We note from tertiary logic that the background of uncertainty is certainty, which is binary logic. The premise of Objectivism is certainty as is the premise of classical science. Certainty as foreground only represents a duality. The "stereo" non duality of the Tao is represented by the yin and yang forces. The yin is the female, passive force while the yang is the male, aggressive force. Yin and yang are constantly interactive and contain small subsets of each other. In tertiary logic, all contrast is with respect to uncertainty (i.e., 0) so that uncertainty forms the background state. As a consequence there is no co existence of opposites and no destructive exo-energetic reaction. It is clear that the state of uncertainty represents zero energy plus one bit of information, while the binary states $(-1,+1)$ represent one unit of energy plus zero bits of information. One bit of information represents one bit of energy in the sense that information is proportional to energy. This corresponds metaphorically to the vacuum state quantum mechanical energy. It is in this sense that uncertainty is creative. From $(+1,0,-1)$ it is clear that background energy is minimal. This is consistent with the fact that meditation is primarily background thought. We make meta-dual thought by adding background to our predominantly foreground thought. Background comes largely in the form of meditation. However, we create background in the fundamental construction of the tertiary logical state of uncertainty. Thus, using meditation is a preliminary process. We must drive into our subconscious mind the possibility of uncertainty as a logical choice. This is the final method of creating meta-dual thought. Methods of certainty as in Western thought consist of foreground thought and are consequently dual. It is no surprise that binary logic is a Western invention. In contrast, we point to the Eastern thought which consists primarily of background thought (e.g., meditation). Rather than representing uncertainty as a separate logical state, Buddhism simulates uncertainty by choosing the center of extremes. Clearly, Buddhism injects background thought. In addition, as mentioned earlier,

Buddhism modifies emotional response. We choose to modify emotional responses entirely.

Just as emptiness can be obscured into nihilism, uncertainty can be obscured into anxiety. In moments of clarity, uncertainty or emptiness becomes openness. For one trained in binary logic there is no clear place for uncertainty, so it can be assigned into different channels of meaning. There is apparently a period of adjustment during which a new logical channel can be established. Since meta-dual thinking sustains both background and foreground thought, there is twice as much thought energy used. In situations in which energy in the form of information is absorbed, thought must represent the energy. If the thought is not powerful enough to represent the energy, results can be destructive. In such cases, it behooves the individual to adopt meta-dual thinking, since it is capable of supporting twice the energy of normal dual thought. Two thoughts in place of one implies that if information flow is treated as energy we get twice the energy consumption. In duality the aim is to select one out of two states. There is a problem with this in that states don't exist as individuals. The result is confusion or incoherence and the states interact, releasing energy. Through the use of tertiary logic, meta-duality selects two out of three states, including uncertainty or the zero state. In this sense contrast is achieved, but there is no interaction of opposites, no incoherence and thus, no release of energy. Based on the use of meditation, meta-dual thinking is extrinsic since we combine separate thought trains. True, intrinsic meta-dual thinking consists of a single train of thought. It is truly "stereo" thinking. The clarity of the initial experience of thought derives from having an extreme emotion coupled with logic. Clearly the machinery for simultaneous thought is right and left brain function. So, the challenge is to train the brain to work in "stereo". The hemispheres have certain tendencies. The right brain tends to input while the left brain tends to output. The right brain tends to be philosophical while the left tends to be mathematical. The right tends to be emotional while the left tends to be logical. An objective is to train each brain to act independently and then learn to work together. There is a larger competition at work in the individual between the conscious and the unconscious mind. Refer to the "unconscious" as the "subconscious" without loss of generality. The subconscious is responsible for 90% of the brain's function. The residual 10% is attributed to the conscious brain. Conscious thought is thus only 10% of the brain's total processing. However, when thoughts are doubled the conscious brain uses 20% of the brain's total processing. This reduces the subconscious percentage to 80%. Consequently, the ratio changes from 10/90 to 20/80. This brings to mind what excess thought energy does to a typical brain. First, output related thought is accelerated. This can be described as mania. Second, input related thought is accelerated. This can be described as depression. Because the excess thought energy occurs in one hemisphere at at time, the individual brains are overstressed separately. The normal way to achieve balance mentally is to participate in a variety of activities which stimulate both brains. However, the most abstract combination of activities is simply reading for the right and writing for the left brain. By focusing on these activities we train the hemispheres in tandom since reading and writing are dual and heavily interactive. The way to have two simultaneous thoughts is to have one passive thought combined with one active thought. In this case, input and output is nearly simultaneous. The subconscious is represented by uncertainty while the conscious brain reflects "yes" and "no" states. In this sense, the brain reflects the full gamut of tertiary logic. The idea for a meditative signal from the left brain first came to me as the concept of air conditioning which cools off an overheated system while dissipating energy. In thermodynamics, an air conditioning unit uses energy provided by the environment to drive energy from a cool reservoir. The air conditioner is the perfect solution to excess thought energy since it dissipates the energy while the meditative thought "music" sooths and thus cools on an emotional level. Since the mechanism was developed by my subconscious, this event provides direct evidence of the ability of the conscious mind to program the subconscious. A conscious thought, if properly prepared and presented will become a reality in the brain. This is a more realistic version of the law of attraction in which the universe becomes the body. The law of attraction proposes that thoughts can attract reality. In order for a thought to become a reality it has to be completely unambiguous and functional. If the thought is inconsistent it will not manifest. There are similarities of this law to neurolinguistic programming which is an alternative method for programming the brain with language.

At this point we return to the discussion of the absolute state of reality. Buddhism identifies emptiness as the real fundamental state. To some extent, this state is consistent with science and the idea of a creative vacuum, not to mention the fact that the atom is primarily empty space. However, consistent with the complex number we acknowledge that reality consists of a real part and an imaginary part which are irreducible components. From the point of view of complexity, reality is not empty. If we identify the imaginary part of reality with quantum uncertainty, then the energy of the vacuum, due to the uncertainty principle, is consistent with the fact that space is not empty. It contains energy.

Key Individuals

There are nineteen key historical players who have implicitly contributed their ideas to this work. I would like to briefly present and discuss them here. Kant said that reason was incomplete and cannot explain all phenomena or prove the existence of God. He believed that only faith can attend to the spiritual world. Plato believed in a world of ideal forms like the circle and the square which cannot be perfectly represented in reality. Socrates believed true philosophy was to question life. Descartes believed in the duality of mind and body. He did not believe that the mind was the brain. He also believed that the evidence of existence was thinking. Nietzsche believed that the fundamental element is the will and that only certain few had the will to dominate. Sartre believed that the world was arbitrary and that the individual had to create a life. He believed that there is existence before essence or that we are not born with a soul. It is interesting to note that this is largely consistent with Buddhism. Wittgenstein believed that philosophy was a game of language and that there are some things of which one should not speak. Kierkegaard was similar to Sartre although he believed in the power of faith. He was also similar to Kant. Bohr believed that reality is determined by interaction. He also believed that all aspects of life consisted of complimentary components like position and momentum. Schrodinger believed that we can only have a statistical understanding of reality. He showed that sometimes objects act like waves and sometimes they acted like particles. Einstein felt that reality was fundamentally deterministic and that quantum mechanics was somehow incomplete. His challenges of quantum mechanics led to some of its strangest properties like quantum entanglement. There is evidence that both Einstein and Newton were searching for a unified theory of physics. Bohm felt that physics should be holistic. He was able to develop a holistic version of quantum mechanics called the implicate order. To some extent, Bohm's ideas are like the holographic view of the universe in which every object is self-similar to the entire universe. Wheeler thinks that the universe is composed of information: it from bit. The universe depends on what questions are asked of it. Occam's Razor is the principle that the simplest solution is the best. Francis Bacon said that knowledge is power. Machiavelli said the ends justify the means. Bertrand Russel believed that religion was a cause of bad things. The Dalai Lama believes that there is a karmic force in the universe. There are consequences to our actions. The combination of these various beliefs creates a coherent basis for my inner universe. As historically important figures, these individuals presented ideas which are important to the outer universe. Therefore, the construction of a unifying concept which incorporates these ideas leads one to an efficient and coherent paradigm for processing information. We might argue that Wheeler's idea of information being fundamental is the strongest component of the worldview presented here. Everything is essentially made up of information. It follows that information extends beyond the physical world into the imaginary world. As we find the concept of imaginary existence in quantum mechanics, information is a natural tool in the description of quantum mechanics.

We point out that there are many more contributors to the ideas in my inner universe and such will be the case for anyone. However, just as a basis space of vectors can represent any vector, a basis space of conceptual contributors can represent any idea. Thus, it is sufficient to focus on a finite number of contributors when constructing a personal paradigm.

Information Engines

There are three essential processes which can be used to interact with and process information: observation, thinking, and acting (OTA). The logical ordering of these processes is OTA. That is we observe a phenomena, we think about it and then we act upon our thoughts. There are other permutations of these processes which we can consider. The first is ATO. That is we act first, then we think about the consequences of our actions, and finally we observe our actions and prepare ourselves for new action. The second permutation is OAT. That is we observe a phenomena, we then act based on our observations and finally we think about our actions. The third permutation is AOT. That is we first act, then we observe the consequences of our actions, and finally we think about our observations. The fourth permutation is TOA. That is we think about some arbitrary phenomena. We then observe the model of the phenomena which we have constructed. Finally, we act based on our observations. The fifth permutation is TAO. That is we think about an arbitrary phenomena, then we act based on our thoughts, and finally we observe our actions. We refer to these configurations as information engines. These engines have a hierarchy in terms of evolution. The lowest engine is AOT, in which actions are primary and thoughts are considered as a last step. The next level is ATO in which actions are primary and thinking is secondary. The next level is OAT, in which observation is primary and thoughts are considered as a last step. The next level is OTA in which observation is primary and actions are considered as a last step. The next level is TAO in which thinking is primary and observations are considered as a last step. The final level is TOA in which thinking is primary and actions are considered as a last step. Consequently, the first and last engines in terms of functionality are reverse permutations of each other. The lowest is AOT. The highest is TOA. The AOT engine prioritizes representations while the TOA engine prioritizes concepts. Concepts and representations are complimentary. That is one concept generates many representations while one representation generates many concepts. We live in a universe which acts to consolidate concepts. This is the primary motivation for the search for a theory of everything in physics and for general solutions to problems in all areas. In this sense, concepts are fundamental. We can therefore say that the TOA engine is the highest evolved information engine. Based on the structure of the TOA, it is clear that it must find representations to match its concepts, while an AOT engine must find concepts to match its representations. It is clear that the universe with all of its representations seems more suited to an AOT engine. Nonetheless a TOA engine is the more powerful since it acts directly on concepts. We should aspire to become TOA engines if we want to deal more aptly with an information rich environment. An information engine sends information from the higher pole of a duality to the lower pole of duality. This process is driven from the natural reduction of a duality. That is, high reduces to low, hot reduces to cold, white reduces to black, and so forth. We might refer to this process as duality diffusion. The transition from one pole to another generates energy which can be output into the brain's environment. An information engine is how the brain transforms information rate into energy. It is clear from this definition that the AOT engine converts energy into information rate, while the TOA engine converts information rate into energy. In an information rich environment it is obviously more profitable to have the ability to convert the available information into energy rather than to have an engine which simply adds to the already abundant information within the environment. In addition it follows that a TOA engine can identify the energy which is produced and thus maintain organization. Processes which rapidly convert information rates into energy may be disorganized if the brain cannot distinguish the energy products. The TOA engine controls the process of information rate conversion and is thus analogous to nuclear fusion as opposed to an out of control process like nuclear fission.

Safety and Freedom

It was Niels Bohr, one of the founders of quantum mechanics, who espoused the concept of complimentarity. The fact that quantities like position and momentum or energy and time are complimentary in physics suggested to Bohr that complimentarity was a fundamental aspect of reality. We find that we can apply complimentarity to safety and freedom, so that if we say that safety=1/freedom, then in a given situation the more free we are, the less safe we are and vice versa. A simple illustration is an animal in the wild. If that animal is placed in a cage, then it is safe from interaction with other wild animals. However, it is not free. Subsequently, if we remove the cage, the animal becomes free, but is no longer safe. Think of the case of the average working person. While at work, they are constrained to do working activities. The compensation for their work provides them with the money to pay for basic needs, like food, shelter, and clothing. Consequently, it is due to their work that they are safe from the elements, starvation and their shelter provides protection from interactions with others. However, because they are constrained to do the activities of work, they are not free. The harder the work, the more constrained they are and the more compensated and thus protected they are from hardship. On the other hand, if they lose their job, they are free of the necessary work activities. However, there is also no longer any compensation, so they are unable to supply the basic needs that keeps them safe from harm. The complimentarity of safety and freedom applies in other cases. Think of the case of extreme experiences or thrill seekers. Some people feel that they aren't alive until they have climbed a mountain or sky-dived. The reason for this phenomena is that freedom is maximized when safety is minimized. That's why people feel free and consequently, alive at the edge of death. One might argue that many feel free and safe at the same time and this is true. Individuals in a healthy retirement situation can feel very free and safe simultaneously. We can conclude that the rule safety=1/freedom is statistical. It is very much like the second law of thermodynamics which says that within an isolated system entropy will never decrease. In fact, we find that entropy tends to increase in most situations. However, there are cases in which entropy remains static. These are so-called reversible states. Likewise, while in most cases, safety=1/freedom, there are cases in which safety is independent of freedom or they may in fact be correlated. Recall the discussion of rational causalities. A rational causality is a sequence of events denoted by a sequence of probabilities and conditional probabilities. Typically, a rational causality resists change. However, when two rational causalities interact, there is the opportunity for change due to the exchange of information. However, in the case for which the probabilities in a sequence are very small and thus the information rate of the associated rational causality is very large, then interaction with a "normal" rational causality will not significantly alter the "larger" rational causality. If we redefine freedom as "subject to interference" and safety as "not subject to interference" then we can write not subject to interference=1/subject to interference. This is an equivalent statement to safety=1/freedom. Consequently, if a large rational causality is effectively unchanged due to interference effects, then it is effectively safe and free simultaneously. We can conclude that a person in a "quantum mechanical" state is simultaneously safe and free.

Are there other complimentary relationships which lie outside the typical scope of physics. We have already shown that learning=1/information. Thus, as information decreases, learning increases and vice versa. Since information is proportional to energy and energy is complimentary to time, we conclude that learning is proportional to time. This follows from experience. The more time we have to interact with a subject, the more we are able to learn about the subject. We can conclude that that there is much to learn about everyday experiences by exploiting ideas normally reserved to physics and mathematics like complimentarity.

Self-Development

Communication is proportional to intelligence. Since, in general the self is divided, inner communication occurs between the different modes of self. When we reach the point at which communication between the modes of self unite the self, we are prepared to direct our communication to the environment. In short, we must first get our inner universe in order before we are properly prepared to communicate with the outer universe. Unfortunately, life doesn't happen that way. Communication processes which are internal and external happen throughout our life. This just means that at some point we must largely isolate our selves and produce a sufficient inner communication process. This is consistent with Maslow's hierarchy of needs which includes as the next to last stage self-actualization. The act of communication with the environment is consistent with the last step of self-trancendence. Ultimately, it doesn't matter what our message is. It just matters that our message is coherent. Whatever we have to say to the environment, it will be consistent with the stable solution we have discovered in our internal journey. The reason why a solution to the self normalizes itself to a solution to the external universe is because our internal information is necessarily consistent with the information within the external universe. Wittgenstein stated that there is no such thing as a private language. The language we use for our own personal communication is therefore a public one.

The privacy we experience lies in the encoding of our personal information. Information encoding is important as it creates a level of isolation so that personal communication can occur without interference from the environment. Therefore, we promote the encoding of information and find it to be consistent with privacy. The fact is that ultimately, we are public since information can be observed and leaves evidence of its presence.

Probability and Intelligence

Based on results presented in this theory, intelligence can distort probability. That is, unlikely events are prone to occur in the presence of intelligence. This suggests that intelligence produces a type of probability field. Perturbations in intelligence produce perturbations in probability. To the extent that energy is mass as shown by Einstein's equation for the equivalence between mass and energy, $E = mc^2$ and since information rate is proportional to energy, we find that information rate is also proportional to mass. There is a surprise factor in information which is metaphorically equivalent to inertia. The greater the information, the greater the surprise and thus, the greater the inertia. Since mass is a property of inertia, in this sense we see that information rate has mass.

(More specifically, we find that information rate has momentum.) This result shows equivalence to the general theory of relativity which states that mass curves space. Intelligence also curves space. The existence of intelligence and curved space is mutual. Some would extrapolate that the Big Bang was due to a singularity in intelligence. This would paint a consistent picture of reality. When we observe the occurrence of accidents, we realize that these are consistent with probability distortions. These probability distortions can thus be associated with distortions in intelligence. Mechanisms for the distortion of intelligence should be rare since intelligence is generally an invariant quantity. One mechanism already described in this work is the speed of thought. The apparent speed of thought is directly related to the apparent distance that thought travels. We have remarked that this can be a complex pathway in the brain system. This is why different people arrive at the same mental conclusions at different rates. Due to the nature of rational causalities, we find that interactions can cause distortions in intelligence since these interactions can cause distortions in probability sequences. Consequently, a large system of interacting individual rational causalities will naturally create distortions in probability. Thus, we can consider the relationship of the frequency of accidents and the system of associated individual intelligences. Such accidents could point to instabilities in the net rational causality. This makes it imperative to create stable and coherent inner universes which align with the outer universe. The program of information compression presented here aims to reduce information congestion and make information transfer a more fluid process. Using the analogy of the two wheels in a communication process, we see that friction is an important component. Without static friction between the two wheels, communication would become distorted. Thus, there must be sufficient coupling in the form of friction between two communicating intelligences (i.e., rational causalities). It is not enough to have a system of substantial rational causalities. It is also important that they communicate. Without communication, effective intelligence is reduced. This fact points to a natural interdependence which is needed in a system of rational causalities. This follows since rational causalities make up the environment of a single rational causality. It is certainly a mutually interdependent situation. Consequently, the ultimate stability of one system depends on the stability of all systems. A system of individuals must rely on each other. This points to a holistic picture of reality. It is consistent with ideas like the implicate order presented by David Bohm. The implicate order is an undifferentiated reality created by quantum mechanics in which systems at different scales are self-similar. The implicate order naturally prescribes a holistic reality. Bohm felt that the successful evolution of reality requires a mutual interdependence. We agree with that viewpoint and our work here is consistent with it.

The Information Spectrum

We have discussed how the frequencies of thought range from the lower frequency of instinct to the moderate frequency of symbols to the higher frequency of emotion. If we correlate this to the electromagnetic spectrum, we find that the visible portion corresponds to symbolic thought. Emotion, as a form of thought is of higher frequency, while instinct is of lower frequency than symbolic thought. Interestingly we find that the information spectrum is inverse to the electromagnetic spectrum. That is, emotions correspond to heat energy or infrared which is lower frequency energy while instinct correspond to ultraviolet which is higher frequency energy. Since information is derived from random sequences in which the information content is proportional to the level of unpredictability, then it follows that higher information such as emotional thought corresponds to heat. In contrast, the more organized energy of generally higher frequency is more deterministic and thus corresponds to lower frequencies of thought. In short, lower information frequencies correspond to higher energy frequencies

Generalized Problem Solving

I am evaluating my work and considering possible applications. One thing that was not discussed was the relationship of information compression to problem solving. It is likely that since problems are biased by duality, that if problems are actively combined, they will destructively interfere leaving a residual of one problem. Consequently, we can say that a collection of many problems will reduce either into one problem or into the null problem. When considering the philosophical influences by key individuals to this work, it is found that there are pairs of complementary problems. For example the worldviews of Sartre and Einstein are essentially opposite since Einstein believed in a fundamental structure while Sartre believed in a fundamental lack of structure. To that extent, they counterbalance each other and destructively interfere. Another example is that of Kierkegaard who believed in faith versus Descartes who believed in logic. We can continue this process of pairing opposites until at last we find Wheeler's concept of the fundamental nature of information. This became the theme of this work. We note that if the cluster of problems is even, the net result is a small residual, while if the cluster of problems is odd, as in this case, the net result is a large residual, which in this case is the solution to the cluster of problems. A problem is an impediment for a state change from state A to state B. We can diagram a problem simply as

$$A \mapsto B. \tag{95}$$

When we solve the problem we arrive at the configuration

$$A \to B. \tag{96}$$

The solution of the problem is the change Δ from A to B. We denote this as the equation

$$A + \Delta = B. \tag{97}$$

It is important that we apply our available resources, R, to the problem. We must have

$$R \geq \Delta. \tag{98}$$

That is, we must have sufficient resources to solve the problem. In the eventuality that our resources are insufficient, we must divide the problem into N reduced problems. In that case, we only need to satisfy the following requirement,

$$R \geq \frac{\Delta}{N} + \alpha(N), \tag{99}$$

where $\frac{\Delta}{N}$ is the divided solution and $\alpha(N)$ is the cost of division of the solution. $\alpha(N)$ is equivalent to interest paid on a loan in which the debt is divided into N parts. If we have a cluster of N problems, then we propose that the solution is divided into N/2 parts when the problems are allowed to interact. In this case the equation for resources becomes

$$R \geq \frac{2\Delta}{N} + \alpha(N/2). \tag{100}$$

Consequently, the division is advantageous for N>2 and if the problems are randomly diverse. If there are at least 2 problems in the cluster, then the necessary resources needed to solve the problem is reduced. Of course, this also depends on the cost of division. We might imagine the cost of division to be equivalent to the information necessary to describe the problems individually. When this reasoning is applied to debt, we see that cost is due to division of the debt into equal reduced payments. This cost is determined as the value of the transaction and thus, the cost depends on the determination of value. In terms of an arbitrary transaction which is part of the solution to a problem, value is arbitrary. However, we might imagine that value is naturally a minimum when the total information involved in the transaction is conserved. It follows that cost reflects the summation of individual quantities of information. The operations of reduction and induction have their own costs, so we imagine that

$$\alpha = reduction + induction. \tag{101}$$

83

Reduction is the information required to reduce each problem into its component parts, while induction is the information required to reconstruct the components into a single whole. It is clear that in order to solve a problem we must add energy or information (rate) to the system we are studying. The problem doesn't solve itself. However, the solution requires induction as a synthesizing component. In that sense, induction is a key component of the solution process. From that point of view, we might value induction more than reduction, such that

induction > reduction. $\qquad$ (102)

Consequently, the cost can be simplified:

$\alpha \sim induction$. $\qquad$ (103)

Referring to the analogy of the debt problem situation, we find that the challenge is primarily to accumulate funds for the repayment of the debt. The problem of acquiring the loan to pay off the debt is relatively minor in comparison. We find from (100) that when the number of problems in a cluster is large, the necessary resources become equivalent to the cost.

$R \geq \alpha(N/2), N \gg 1$. $\qquad$ (104)

Thus, the required cost to solve the problem originally consisting of a cluster of problems becomes the cost of induction of the components of a single problem into a single solution. Since the solution is induction, it must necessarily have the form of a general rule. Consequently, the solution becomes general and not specific. A general solution requires less information than a specific solution.

In the example of this work, the net solution is the general statement that information is the primary material of the universe which comes from Wheeler's statement, "it from bit". Our presentation is consistent with this general statement which we make specific by comparing information flow to electrical current. This recognizes the electromagnetic field as a primary component of reality. That is, the universe consists of energy. This energy is largely contained in fields. The electromagnetic field is likely the primary field in reality. Noting the fact that the intelligence equation which is part of the solution relates directly to spatial curvature as does gravity, there is a clear connection to the gravitational field as well. This situation suggests that problems should be considered collectively rather than separately. The combination of problems presents a natural division of the solution such that finite resources are sufficient to solve problems.

Economics

Economics facilitates the flow of money. To the extent that money can be an element of communication, economics facilitates communication. Money is a primary resource. A resource is something which drives or enables an action. Communication, in this sense is the exchange of money. Money is equivalent to energy in the sense that it enables action. More specifically, the energy of money is how often the money is received. High energy money is received often, while low energy money is received infrequently. The transformation of mass into energy is the underlying process for the increase in money. Consequently, money increases at the loss of organized mass structures. Since entropy increases with time, we might conclude that the increase in money comes with the necessary increase in entropy. The statement, "time is money" is false, since time and energy are complimentary. The statement only makes sense from the point of view that money is acquired over time. The problem is that specific intervals of time do not correlate to specific amounts of money. At this point it is beneficial to discuss complimentarity. Consider a sine wave of a specific frequency. The extent of the sine wave is infinite. Consequently, it is infinitely uncertain in time. However, since it has one frequency, it is infinitely defined in frequency or energy. Now, consider a single impulse in time. It is by definition well defined in time. It consequently is broad in frequency or energy content. Now, let us compare this to money. Consider a monetary sum of one hundred dollars which is represented with one denomination, say one dollar bills. We will suppose that the money is located in one place. Consequently, the money represents different times since we will assume that before the money was received as a lump sum, different portions of the money was acquired at different times. Now consider another monetary sum which is received at regular intervals. Suppose that there are ten dollar payments received every month. For the first monetary sum, there is time certainty, but the frequency of payment varies. Thus, we have time certainty, but energy uncertainty. However, in the second monetary sum, there is time uncertainty since the payments are spread over time, while there is one regular frequency of monthly payment. Thus, there is time uncertainty, but energy certainty. From these examples, we understand the uncertainty relationship between time and money. Since energy comes in two essential forms, that is, potential energy and kinetic energy, we surmise that money also comes in two essential forms, real money and potential money in which we have substituted the word "real" for kinetic. Real money is money that exists. Potential money is money that potentially exists. Systems like banks deal in both real and potential money. For example, cash is real money. On the other hand, debt is potential money to the person owed. An investment is also potential money. When potential money becomes real, an economic situation reaches fruition, since the final goal of an economic venture is always real money. Money as energy can do work on financial systems, causing fluctuations in their net worth. Variations in the stock market which consists of mutual shares of companies is a reflection of changes in potential money because of financial interactions. It is because potential money is intrinsically tied to intention that the stock market is subject to spurious variations. These spurious variations reflect the change in intentions of members of the system. Economics, as a result of the effects of spurious energy variations, can be very complex in a system of interacting knowledge networks (i.e., rational causalities). A primary rule to consider is that if money is considered to be a resource and goods are considered to be a resource, then in isolated systems the net resources are conserved. The goal of any economic system is to have efficient communication of money. Consequently, the theory presented in this work should be consistent with successful economics.

Enlightenment

It is worth discussing the possible relationship of this work to the concept of enlightenment. When we refer to enlightenment, we are not referring here to the eighteenth century movement based on reason and individual rights. Enlightenment is typically associated with spiritual attainment. It is, however, an ambiguous term. It is particularly relevant to Buddhism, in which it is the primary goal. We ask ourselves, what does enlightenment really mean and how is it pertinent to this discussion? In Buddhism, the ultimate problem is human suffering. The attainment of enlightenment marks the end of suffering for the individual who becomes enlightened. It is consistent with the concept of supreme knowledge and wisdom. It has other meanings however. The author of the Power of Now, Eckhart Tolle, claims to have reached enlightenment when his time sense became consistent with a sense of now as the meaning of time experience. Time is sharply defined in this case. We know from physics that when time is sharply defined, energy or identity is uncertain. To some extent this is consistent with the lack of self promoted by Buddhism. Logically, we ascertain that time certainty and Buddhist enlightenment are identical states. We might say that this state is consistent with an eigenstate of time. Although, as previously stated, there is no such thing as an eigenstate of time in the physical sense. Consequently, living in an eigenstate of time is a purely imaginary existence. To the extent that those, who are enlightened in this way, have no sense of self, because they have perfect location information, they have no personal bias for knowledge. Thus, they have a natural objectivity which is superior to the typically biased perspective. We might thus conclude that enlightenment means perfect objective knowledge, and we naturally transcend the self, since we no longer have a self. This is consistent with the last state in Maslow's hierarchy of needs, self-transcendence.

The major proposition of this work is to develop an eigenstate of energy, which is consistent with uncertain time. When time is uncertain, the subjective sense of its progression is accelerated relative to a traditional time sense. Having an eigenstate of energy in this sense dilates intelligence according to the axiom that intelligence is fundamentally proportional to length. This state is naturally a high intelligence state, and as such, it is consistent with the concept of enlightenment. From this general perspective, it seems that enlightenment is consistent with existing in a well-defined state. Consequently, we might argue that enlightenment is consistent with having a highly defined state of existence. In other words, enlightenment is consistent with the concept of coherence. Coherence is presented in this work as the primary solution to congested knowledge networks. It is worth mentioning the proclaimed experience of Ken Wilber who professes to live in a constant state of non dual awareness. Going beyond duality is consistent with a state of enlightenment. The energy state discussed in this work is a meta-dual state, consistent with enlightenment. The particular nature of Wilber's state is likely compatible with the concept of meta-duality, because in both cases, experience is extended beyond dualism.

Thought as Information

We can examine the nature of thought in light of the newly presented information physics. Since thought is composed of information, it is measurable as a unit of length. Longer thoughts have more information. Shorter thoughts have less information. Thoughts have momentum. This is reflected in the intertial nature of information. Information presents surprise when encountered. Information is the reduction of a priori uncertainty. The surprise factor is consistent with momentum. Since thoughts move, they have momentum or velocity. As such thoughts represent information rates. In this sense thoughts are an integral part of communication. When thoughts manifest, they become words. Thoughts are concept and words are the associated representation. The energy for complex superluminal thoughts was given earlier as $E = inp_0c, p_0 \in real\ numbers$. From this perspective, the energy of superluminal thoughts are discretized. In a consistent fashion, we can see that superluminal thought is dilated while time is contracted such that there is an increase in information rate and thus intelligence as follows.

$$I_n = I_0(1+n^2).$$
(105)

We must also consider the case for which thought is not apparently super-luminal. The average velocity for an interval between points A and B is the interval AB divided by the time required to traverse the path. We can write this as

$$\bar{v} = \frac{AB}{\Delta t}.$$
(106)

If the thought meanders and travels a circuitous path between A and B, then the time interval is augmented. Consequently, the average velocity decreases. The momentum of the thought will accordingly decrease. We will suppose that the division of momentum is by the factor μ. The energy will thus be divided:

$$E \to \frac{inp_0 c}{\mu} .$$

(107)

Consequently, thought energy is reduced when thought motion is inefficient. We imagine that this occurs when the thinking process is inefficient. We can conclude this discussion with the point that thought energy is a measure of intelligence and thus, intelligence is reduced when thoughts are inefficient.

There is an additional factor of communication which we would like to discuss. When information is transmitted there is the collapse of a priori surprise and the information is measured in bits. We propose that there is an additional factor to information transmission which is proportional to the engagement or coupling of the information between the transmitting agency and the receiving agency. Thinking of the analogy of two wheels rotating in contact, we imagine the coupling as proportional to the force of contact between the two wheels. If the wheels are in contact and rotate together without slippage, then there must be sufficient static friction. The frictional force is proportional the normal force at the point of contact. We say that the coupling between the two wheels is proportional to the force of static friction. Thus the information transmitted has another factor which compares to the coefficient of static friction. Consequently, the transmitted information for the first transmitter can be written as

$$I_1 = \frac{-k_s \log(p_1)\omega_1}{\mu_0 c},$$

(108)

where k_s is the coefficient of static friction between the surface of the transmitter and the surface of the receiver. The coefficient is a measure of the efficiency of the transmission.

This modifies the intelligence transmission for superluminal thoughts to

$$I_n = k_s I_0 (1 + n^2) .$$

(109)

Typical ranges for the coefficient of static friction are given by

$$0.5 \le k_s \le 0.8 ,$$

(110)

which translates into an efficiency between 50% and 80%. A realistic application of this equation is very interesting. If we interpret I_0 the length as being equivalent to intelligence quotient, I.Q., that is

$$I_0 = I.Q., \tag{111}$$

then for an I.Q. of 100 points and a superluminal quantum number of n=1, we have

$$I_1 \rightarrow k_s 200 \text{ or}$$
$$I_1 \rightarrow 100,160. \tag{112}$$

Thus, we see the potential augmentation which occurs due to superluminal thoughts and the specified range of coupling.

Information and Time

Based on quantum mechanics time is not an observable quantity. Although we readily make time measurements, based on quantum mechanics time measurements are not valid as observable quantities. An example of an observable quantity is momentum. The non-observable nature of time is somewhat controversial. It was first explained by Pauli as a truth because time is represented by the spectrum of energy and there is no such thing as negative kinetic energy. The problem is explained based upon the confusion of complimentary information states of location and identity. When the location of information is known, the identity is unknown. When the location of the information is unknown, the identity is known. When we make a time measurement we can think of it as measuring the length of arc along a circle whose circumference makes up the clock. If we think of the clock as having a linear mass density, then a length of arc is equivalent to a mass measurement. Consequently, time measurement is a mass measurement. Since mass is energy, mass measurement is energy measurement. Thus, a time measurement is equivalent to a measurement of energy. That is identically the reason why time is not observable. We can contrast this to position and momentum which are complimentary. A position measurement is also equivalent to a mass measurement if the length measured has a linear mass density. However, momentum is not equivalent to energy because of the velocity variable. Consequently, a position measurement does not confuse a momentum measurement. Based on the random phase selection of a clock, information is a time measurement that is actually a measurement of energy. Thus, information rates are measures of time per unit cycle of the clock and they are in fact energy measures.

Knowledge is energy
The equality of information rate to energy is an important correction of the statement that information is equivalent to

energy. This follows from the fact that energy is conserved. Since energy is conserved, it is precisely correct to say that information rate is conserved. Obviously, this must refer to a net information rate. Consequently, we say that the net information rate is a constant in the universe. This is equivalent to saying that communication is a conserved quantity. This follows since it is clear that information itself is an ever-growing commodity. It is thus, wrong to say that information is conserved. It is correct to say that information rate is conserved. This implies that while information continues to grow, the average rate of communication is a constant. Based on the definition of intelligence as being proportional to communication, we can conclude that intelligence is conserved. Consequently, it is clear that intelligence is not equivalent to knowledge. Knowledge is ever changing and in particular, knowledge is believed to be growing. Knowledge is the currency of intelligence. The statement knowledge is power leads to the conclusion that information is energy. This is, in fact, erroneous. The correct statement is that knowledge is energy, because for a given quantity of knowledge, that knowledge is proportional to information rate.

God

The existence of God is a controversial subject. There have been no overwhelming proofs or evidence either of the existence of God or the non-existence of God. A large amount of scientific study is able to avoid discussions of God. It is only at the boundaries of physics that God has been discussed. For example, it is widely believed that the universe originated in the Big Bang approximately 13.7 billion years ago. The Big Bang is considered not only to be the origin of the universe, but the origin of time and space. The question of the cause of the creation of the universe naturally connects to the existence of God. As a consequence, the cosmos is pertinent to God. At the other end of the scale, there is the existence of quantum mechanics in the microscopic world. Quantum mechanics points to an indeterminate universe. Consequently, the question of God might also arise in quantum mechanics. Since the essential topic of this work is intelligence, we are able to discuss maximal and minimal intelligence using the tools and equations presented. Consequently, it is natural to question the nature of infinite intelligence, since infinite intelligence is an attribute of God. This work is naturally pertinent to God. In fact, it follows directly from the continuity of this work that infinite intelligence exists if frequency and probability are not coupled. Thus, God is pertinent to this work. I believe that it is not a question of being pro-God or anti-God. It is rather, a question of pursuing the most pertinent truth. When discussing a consistent model of reality, we should be ready to pursue whatever idea or concept that arises. There are two interesting formats for infinity represented in this work. The idea of a complex meta-duality with an infinite yin as emptiness and and infinite yang as God is one of the formats presented. In this yin and yang structure, the yin is the real part and the yang is the imaginary part. Since this structure is complex, it is irreducible. To some extent this meta-duality combines the notions of Buddhism with emptiness and Christianity with God. Emptiness is the final real notion, while God is the final imaginary notion. The fact that complex numbers have a real measure or modulus of $|x| = \sqrt{a^2 + b^2}$ for the number $x = a + bi$ suggests that in this case, we have $x = 0 + \infty i$ which yields $|x| \to \infty$. Thus, this meta-dual state has infinite measure due to the God component.

This suggests that God has a measurable impact on reality even though God is purely imaginary. The other format for infinity presented is in the intelligence equation $I = \dfrac{-1}{\mu_0} \dfrac{\log p}{r}$. When $r \to 0$, there is infinite curvature as for a singularity. If probability is independent of radius in this case, intelligence becomes infinite. Considering the belief that God represents infinite intelligence, we have a possible real projection of infinity. Of course, it is questionable that the center of a black hole exists in reality since the laws of physics are disrupted there. From this perspective, we might consider that infinite intelligence is never truly within reality. Consider the possibility of infinite intelligence being sourced by bits of information. A pure bit of information has zero radius, since it has a zero minimum size. It is also important to point out that imaginary existence is consistent with both God and thought energy. Thus, we might think of God as infinite thought. Perhaps the Big Bang was initiated as a probability distortion created by infinite thought.

Work

When we work, we exchange energy for resources. If work were totally conservative, then the net energy or energy equivalent resources would be conserved. However, energy is only conserved within an isolated system. The processes which relate to work consist of a set of interacting systems. In any one of those systems, energy is not conserved. It is only when the total system is accounted for that energy is conserved. The fact that energy is not conserved within a particular energy or resource containing system is the reason why some people get rich and others get poor. The fact that the rich and poor exist points to imbalances in energy exchanges. We might say that this points to imbalances in communication. When considering two systems, the interaction is imbalanced when one system transfers more information than another. It is believed that systems like those presented in this work will promote balance. Consequently, the enhancement of intelligence promotes balanced systems of communication. We must emphasize that it stands to reason that there will remain hierarchies in intelligence that will distribute themselves throughout the total system. However, it is in the interest of full expression that the intelligence potential of every individual is reached. Work is meant to be a beneficial process. It is interesting to note that work is a competition of sorts. We exchange work for resources. Since resources are conserved, work is competitive. Nonetheless, it is supposed to be a fair exchange for resources. It is hoped that ideas like those presented here will help to reinforce fair exchange of energy and information.

Play

When we play, we exchange energy for entertainment. Play is an opportunity to enjoy our temporal experience. Imbalances within a system occur when some people play too much and others play too little. Play is complimentary to work. Consequently, when you play hard, you should work hard. Unfortunately, some people don't recognize this complimentarity. We should establish a communication between our experiences of play and our experiences of work. To that extent play and work should be synergetic. It is only by having a balanced and fluid communication that the balance between play and work can be maintained. To the extent that all ideas are part of a duality, play and work form a complete set. Consequently, the arguments presented to describe work essentially describe play.

Efficient Intelligence Interactions

Given two interacting intelligences, the net intelligence is the sum multiplied by the coefficient of static friction since this coefficient is equivalent to the efficiency of interaction between two intelligences:

$$I_{net} = k_s(I_1 + I_2).$$ (113)

This equation can also be applied to the two hemispheres of the brain. The net intelligence of the brain is the sum of the left intelligence and the right intelligence multiplied by the efficiency of communication between the hemispheres. We can assume, for example, without loss of generality that a brain has equal intelligence in both hemispheres. When the person interacts primarily with one hemisphere at a time, only one intelligence is presented. Alternatively, outside interaction can be with both hemispheres if they communicate prior to the interaction. This joint process would entail communication between the hemispheres. For example, if the IQ of a person is 100 points and there is balance between the hemispheres, then for a reasonable efficiency of $k_s = 0.75$, the net intelligence would be $I_{net} = 150$. Thus, intelligence can be augmented by enabling the hemispheres to work together during mental transactions. Partitioning of thought naturally occurs in terms of analysis and synthesis. That is, analysis of the original thought and synthesis of a response. It is natural to allow the left hemisphere to engage in analysis and to let the right brain engage in synthesis. It is interesting to note that if the interaction between the hemispheres is superluminal and not subject to a physical channel, then the efficiency goes to unity: $k_s = 1$. In such a case we would say that the information coupling is perfect.

Energy and Time Eigenstates

Energy and time are complimentary. However, there is evidence that time is a purely imaginary state, while energy is real. Quantum mechanics finds that time is not an observable quantity. Only finite time intervals are observable. Consequently, time instants don't really exist. Nonetheless, we appear to be able to detect time, so we present time as an imaginary construct. In this sense energy and time form a complex ensemble. Energy is the real part and time is the imaginary part. (This is similar to the arrangement for emptiness and God.) Thus, we posit that having a fixed time sense corresponds to an eigenstate of time which is necessarily imaginary, while simultaneously, real energy is fuzzy. This is the state of no self, since identity corresponds to energy. The state of no self corresponds to observing only the present moment. It has been typically correlated with a state of enlightenment. Since it is based on the purely imaginary, we propose that the eigenstate of time may be a meta-stable state. In other words, it may be difficult to sustain such a state. In the interest of promoting coherency, we advise that an eigenstate of energy is a more stable state to focus on, since energy is real. Having an eigenstate of energy means that the self is well-defined and time sense is fuzzy. We suggest in these situations to have an identity which is actually a category rather than a particular. Karmic deficits and benefits will thus be associated with the category rather than a particular identity. For example if the category were the rich, then when there is karmic payment, the rich would suffer a deficit or gain a benefit. This is in contrast to having the identity John Smith, in which case karmic payment would tie to John Smith, the particular person.

Having a fuzzy time sense is consistent with the sense that time is contracted. This result follows directly from relativity applied to complex superluminal velocity. Complex superluminal thought corresponds to enhanced intelligence through the phenomena of length dilation, for which intelligence is proportional to fundamental length. There is coherency in both a temporal eigenstate and an energy eigenstate. We propose that having an energy eigenstate is more sustainable because energy is real, while time is imaginary. As far as the stability of one state versus another is concerned, it is unclear. At first glance it appears that there is no mechanism for switching states, however there may still be ways to switch from an eigenstate of time to an eigenstate of energy and vice versa. This requires that thought speed become variable from causal sub-light speeds to complex superluminal speeds. When thought moves at sub-light speeds, intelligence contracts and time sense expands, but when thought moves at superluminal speeds, intelligence dilates and time sense reduces. The problem with this scenario is that it requires that intelligence vary, but intelligence is by definition a conserved quantity. Enlightenment may be more closely associated with wisdom since it correlates with the most efficient way to be or the most coherent way to be.

Education

Learning is a primary component of information processing and corresponds metaphorically to induction. Induction occurs when one current causes another current or when information flow within one knowledge network causes information flow within another knowledge network. Learning requires a communication process between, minimally, two systems or two knowledge networks. Education is the system which encourages learning within an organization of knowledge networks for the purposes of the growth of information. Education substantially deals with background information and creativity primarily deals with foreground information. From this perspective, education serves as a reference point for knowledge and it allows knowledge to be extended continuously within a cultural framework. The resistance of a current to changes is a measure of Lenz's law. This suggests that there is a sort of work function associated with learning. In other words, in order to learn new material, we must first overcome old meaning and supplant it with new meaning. The more different the new material is from the old material, the more difficult the initial stages of learning becomes. This is consistent (as mentioned earlier) with Thomas Kuhn's work, the Structure of Scientific Revolutions, in which new and old paradigms are seen to be competitive. Education is necessary for the continuity of knowledge. It exploits induction within knowledge networks. Education can be private or public. Public education systems are intrinsic to societies, while private education systems are pursued by small groups of individuals. We propose here that both private and public systems of education are necessary to the development of coherent pictures of information processing.

Faith, Uncertainty and Information

It is found in quantum mechanics that uncertainty is irreducible. This follows from the uncertainty principle which states that one compliment can be relatively known while the other compliment is relatively unknown. The product of uncertainty is thus a minimum. This principle suggests that we can never have perfect knowledge. Faith, however takes us beyond the boundaries of science. Faith is normally in the confines of religion. It suggests that we can know with certainty that which we have no evidence for simply because of the power of our belief. To some extent, it might be argued that every belief is a matter of faith. It could be that a scientific proof or equation is simply a tool of faith meant to enhance our belief. In either case, we might recognize that faith, in its pure form uses no tools other than belief itself. We propose here that when statistics seems to be the final determination, certainty can be obtained by the addition of faith. In fact, we propose the equation $faith - uncerta \text{int} y = certa \text{int} y$. To some extent, the application of the theories presented in this work rely on a measure of faith. We might propose that all theories rely on faith. It is in the corroborating experiment that we find a measure of certainty consistent with our original conviction. It could actually be that belief holds reality together. To the extent that the object of belief is meaning, it follows that meaning drives reality. Since meaning is metaphorically equivalent to voltage and voltage is potential, we must ultimately look forward to new potential in order to be properly motivated. Einstein always felt that quantum mechanics was incomplete. It may be that faith was the missing component.

If it is true that information rate is conserved, then at some point the rate of information discovery will maximize and eventually reduce. It seems that information rate is a finite resource. To the extent that information rate converts to energy and energy converts to mass, the entire universe is essentially moving information. The end of the universe will come when energy is completely entropic or disordered and there is no more useful work to be acquired. This same scenario can be applied to information. Eventually, information becomes entropic and is no longer useful to the learning process. Every resource is finite. Thus, we must find a way to become efficient with information usage. It seems that the intrinsic property of dualism is a natural source of information and thus energy in the universe. Duality as a source of energy is consistent with efficient information engines. An information engine extracts work from the separation in meaning of information and the tendency of the high end of a duality to reduce to the low end of a duality. A heat engine is a subset of a information engine. We can answer the question as to how efficient an information engine is based on the interaction of two engines. When two knowledge networks interact the maximum efficiency of the interaction is about 75%.

When two knowledge networks interact, they share a collective unconsciousness. Maximal communication through this unconscious channel will be through a complex superluminal velocity. Time will contract in this case as follows.

$$T = \frac{T_0}{\sqrt{1+n^2}},$$ where n is the velocity multiple of the speed of light,

c and T_0 is the time interval within the inertial reference frame, which is the reference frame at rest. For the case, n=1, we have

$$T = \frac{T_0}{\sqrt{2}} = 0.707T_0.$$

The efficiency of interaction is proportional to the correlation of time. For instance, if time were infinitely contracted, then the moving clock would be infinitely fast. Consequently, the length of a time interval would be zero and there would be no time for interaction with the inertial reference frame.

Information is not conserved

We have discussed this elsewhere, but we reiterate a few points here.

The choice of information rate as energy contrasts to selecting information as energy. The fact is that information grows as a form of entropy and entropy is not conserved. It is clear that information is not a conserved quantity. However, we posit that information rate is a conserved quantity. We interpret this as meaning that there is a maximal effective information rate in the universe. The rate at which information can be transported from point A to point B depends on the total available information rate. From this perspective, rates can be compounded. This is equivalent to saying that there is a maximal possible transfer rate for information. We posit that this rate is none other than c, the speed of light. The speed of light is a conserved quantity. This result poses no problem for thoughts since thoughts do not move in a traditional sense. Thoughts move at superluminal speed in the same way as the information which is transferred during quantum entanglement. To some extent, we say that thought motion occurs outside of our universe. In that sense, we are interpreting the space of consciousness as being a separate space from the universe.

Reflection on Essential Points

We have presented what is essentially a theory about intelligence and information in the hopes of promoting the efficient use of information. In that sense, we are able to compress information and avoid the pitfalls of information congestion. What this suggests is that as our culture advances and uses more information, we are forced to fit this growing deluge of information into our personal paradigms. Because we have our own personal inner universe which must merge with the common outer universe, we are forced to fit our paradigm into or at least merge it peacefully with the paradigms of the outer universe. Systems of thought such as Christianity and Buddhism have recognized the universal existence of sin or suffering due to the misuse of communication. This has been stated, essentially as the misuse of energy, which is proportional to information. The peaceful coexistence of inner and outer paradigms creates a peaceful environment both for the inner universe and the outer universe. We are hopeful that ideas like the ones presented here eventually may help society in its evolution to a peaceful universal existence or a utopia, if you will. We propose here that the essential information needed to have a peaceful existence is known, but weakly followed. The reason for this is the lack of sufficient will. Will is developed with time and focus. It is something we have intrinsically but I believe it can be enhanced with practice. When we want to make some idea a reality, we need to think about it in a patient, but focused way. The law that tension creates continuity produces the manifestation of ideas into reality. Once an idea is real, its further evolution is promoted by the symmetrical law that continuity creates tension. We make reality and reality makes us. We have stated that reality is fundamentally a duality. That is, reality is a set of compliments or opposites. Because of this, we find that isolated meaning does not exist. One thing is defined relative to another. For example, we understand the meaning of heat only in contrast to the meaning of cold and vice versa. This situation of duality tells us that we essentially need a dual thought process to interpret reality properly. Essentially, we need to consider two things at once. We have used the analogy to stereo sound to explain this situation. Normally, reality is monophonic (i.e., mono). The challenge is to experience reality in stereo. Ideas like the middle path of Buddhism avoid extremes and yield an intermediate representation which might be described as gray as opposed to black and white. In contrast the stereo embraces both extremes and creates a relatively colorful experience. To the extent that the Buddhist experience does not distinguish between extremes, it is considered to be non dual. The stereo experience might be described as meta-dual, a subtly different variant of non duality which extends itself beyond extremes. The reduction of information into dual components which can be considered to be opposites is represented in such examples as light vs. dark, male vs. female, war vs. peace, determinism vs. randomness and so on. The specific identity of charge relates to a form of duality practiced by the Chinese culture called the Tao. The Tao subdivides reality into the yin and the yang, the negative and positive counterparts which interact to produce energy. The subtle difference between the Tao model and dualism is that there is a little yang in yin and a little yin in yang which creates commonality between the charges. In duality, charge is homogenous. The fact of the charged nature of reality brings us to the concept of the metaphorm. A metaphorm is a common underlying concept which is indifferent to specific representations. The idea was first proposed by Todd Siler in his books including, Breaking the Mind Barrier. The metaphorm naturally compresses information by consolidating two or more ideas into a single concept. In the sense that concepts are complimentary to representations, we might say that the metaphorm moves the information from the space of representation to the space of concept. With the property of charge within information, we are able to make the accurate metaphorm of charge flow to information flow. In fact, we have shown that this metaphorm creates an identity or an equivalence relationship. In the work that follows this idea, we have presented extensions of the current to information rate metaphorm to describe other relationships such as voltage to meaning and resistance to probability.

Using these and similar analogies, we were able to derive a metaphoric description of information theory in terms of the physics of electromagnetism. We recognize that the work done is only the beginning of possible extended analogies which would serve to fill in the associated theory. We continued with analogy to arrive at a description of intelligence based on information content. A physical equation for intelligence was presented which is proportional to information content in terms of probability or the self-information and the curvature of space. From this equation, we intuit that intelligence has the potential to warp probability and space. This equation transforms variables from curvature into frequency and at this point we were able to derive a more quantum mechanical equation which is the correlate to quantum mechanical energy. This equation represents the conclusion of the main derivations of this paper. Evidence of the validity of this equation is simply represented in the encoding of symbols which requires curvature of motion for high information rates. An idea presented which is also very important is the idea of a rational causality, a sequence consisting of probabilities with various levels of conditionality. A rational causality is essentially a reality. It may be deterministic or it may be random or some variation thereof, depending on the probability distribution. The essential objective is for a rational causality to remain or become stable.

Unstable rational causalities are models for conditions like mental illness or mental imbalance. Mental imbalances are often the result of overstimulation or understimulation with information. We point out that since information can interfere like waves, constructive and destructive interference can readily generate overstimulated or understimulated states. The work on rational causalities points to the net experience of reality when two realities interact. We find that the remaining probability is that of the larger reality. The proof of that result is worth evaluating here. Consider two events which are essentially two sequences. The experience of event one is reflected in the sequence of probabilities that describe the event (i.e., rational causality). Likewise, the experience of event two is reflected in the sequence of probabilities that describe the event. When we sum the two events or when they interact, there is a new, net event. If the experience of event one is much larger than the experience of event two, then the experience of the net event is approximately equal to the experience of event one. In other words, event two is not relatively detectable. Consequently, the probability distribution of the net event is the very similar to that of event one. Consider the first members of the two original sequences. There will be a first probability of event one and there will be a first probability of event two. Add these events. The net probability will be the product of the two probabilities. For example, if probability one of one is $p_1 = 1/2$ and probability one of two is $p_2 = 1/2$, then the net probability that each event happens is $p_1 p_2 = 1/4$. Now, suppose that the probability one of one is very small, that is $p_1 \sim 0$. Then, the net probability that each event happens is $p_1 p_2 \sim 0$, which approximates the probability one of one. Essentially what is being said here is that when two events are combined, we notice the larger event and if the larger and more unpredictable event is large enough, it will appear as if there is no perturbation to the event. This idea is central to stability. A rational causality that is large enough will tend to be stable. Since an event of sufficiently high complexity is indistinguishable from an event of high entropy and changes in an event tend to increase entropy, a highly complex event will decay slowly. We are thus compelled to have highly complex rational causalities in order to insure stability. It is interesting to note that the attraction of two rational causalities or the repulsion of two rational causalities is magnetic. This follows from the magnetic interaction between two current distributions. We point out that the terminology for a rational causality is the same as for a knowledge network. Each may be described in terms of a sequence of words or a sentence. The number of symbols which represent a knowledge network or a rational causality reflects the size of the system. In that sense, the most elementary systems consist of just a single symbol. It was suggested that anomalous variations in intelligence will correlate with anomalous variations in probability. Based on the intelligence equation, these distortions could couple to physical reality via the curvature of space. Variations in the curvature of space could cause gravitational anomalies or their equivalents. In other words, there could be a real connection between variations in intelligence and variations in the physical world. Distortions in our worldview could correspond to anomalous events occurring within reality. This points to a continuity between the mental and physical universes that extends the apparent operation of consciousness. A major result of this work is the correlation of information to length. In a sense, it would follow if information were made congruent to any fundamental unit of measurement. However the unit of length is consistent with the network of associations built from the fundamental metaphor of current to information flow. Thus, there is an internal consistency to the metaphor of length. This determination is consistent with the well known relationship of heat energy to information as with demonstrated with a discussion of Maxwell's demon. The additional result of correlating length to information is the effect of relativity. Relativistic lengths contract. This points to a reduction in intelligence as velocity increases. However, things change dramatically for complex superluminal velocities. In that case, length dilates (and time contracts). Consequently, we can say that complex superluminal thought is associated with increased intelligence. We ultimately derive an equation for intelligence which is proportional to the level of superluminality. We speculate here that having a superluminal thought is equivalent to what has been considered enlightenment. It is important to note here that while it seems that presented topics vary widely, we are just presenting the consequences of a single metaphorm. In the interest of consistency and coherency, we discuss the worldview as much as possible in this context. Because of this, there is much that will not be discussed here. Such things will have to be left to other related work. One of the results presented here is the existence of a virtual universal clock. We visualize the random selection of phase for such a clock. The phase selection corresponds to a general communication process if two observers are involved or if synchronized clocks are involved. A fundamental, virtual clock is a source of relative comparison which allows physicality to be normalized. There is some indirect reference to the problem of quantum mechanics in this work. We presented the equation, faith-uncertainty=certainty as a way to complement uncertainty to yield certain results. The result is that the compliment of science is faith. To some extent, it may be that scientific and mathematical equations are a sort of consolidated faith. Both faith and science may just be the tools needed to understand and accept reality. This type of idea heals the schism between religion and science and suggests that in order to understand the complete picture of reality we need both. We present in this work the concept of an omega state for brainwave activity. The idea is that the omega state involves meta-dual thought or thought that sees reality in stereo (i.e., the extremes combine to form a higher state). The potential of thought is that its motion has a complex superluminal velocity. When thought motion is superluminal, the brain is able to operate at peak efficiency. The result is an augmented intelligence. Thus, the ultimate result of processing compressed information is an improved communication system which directly enhances intelligence.

This work has been an exploration of the ideas originally presented by Todd Siler. His vision of the advantages of metaphorms was that they would extend our understanding of reality. I offer this work in support of that belief and I hope that the efforts in this direction do not end here.

End Note: Spirituality

The concept of the metaphorm is somewhat mystical. It points to what I have coined as a meta-duality or a world of the realization of extremes. We should experience reality in stereo. Going beyond duality is an exploration. It moves us beyond war and peace to a state that exists beyond either of them. These ideas are meant to create an impetus to move knowledge to a new regime. We have yet to discover what lies beyond the horizon.

The achievement of higher states of consciousness has been a major objective since human culture began. The reasons are varied. One major reason has been the need to alleviate pain and suffering. Other reasons have been on the other extreme. For example, human beings constantly desire power over their environment and over one another. The evolutionary drive of survival of the fittest could be the essential reason for this. Whether used for good or bad, the acquisition of a higher state of consciousness is part of the logical progression for evolution. An essential element of this work is the need to organize the self or the inner universe first as a priority development. We often spend a lot of time trying to change our environments directly in order to improve life for ourselves and others. However, the most pertinent problem is the problem of self. This is consistent with Maslow's hierarchy of needs which lists self-actualization and self-trancendence as the highest goals. Once we master the self, we are then prepared to extend our efforts out into our environments which will consist of others and resources.

This is the essential plan presented here. Using all of the available knowledge to achieve our goals makes sense. In this work we present knowledge about information and intelligence because we feel that they are important components of the arsenal of essential tools for manipulating reality. Knowledge is energy and we need energy in order to solve problems. Since life constantly presents us with problems, we are motivated to accumulate all of the available energy. To some extent life is a zero-sum game. That is, some win at the expense of others losing. That is the nature of competition. Consequently, we owe it to ourselves to gather all of the useful tools that are available to us. Information management as presented in this work is a necessary process for dealing with the environment of congested information. The growth of information is a necessary consequence of advancements in technology and culture. Nonetheless, the pitfalls associated with information congestion must be confronted. There are pros and cons to every situation. We must not only be prepared to embrace the pros, but we must be prepared to negotiate our way through the cons. This work alters the perception of the division between science and spirituality. In the sense of this presentation, there are no divisions. Everything can be viewed as metaphysical in a unified sense. The program is that we simply metaphorm everything into science. Consequently, there is a consilience of knowledge in this process. The rules of science will project into all disciplines and there will be some natural discoveries and extensions of the known laws. In this sense, there is no need for division. Spirituality seems to be a description meant to describe phenomena which lies to the abstract side of science. However, we present here the concept that science is not intrinsically limited. In the end, it doesn't matter what we call it, whether it is science or truth. To the extent that our scientific developments are based on our particular perceptions of reality, we acknowledge that there is relativity to that knowledge. Thus, we say that truth is not absolute, but relative to the universe we live in. This acknowledges the fact that we live in a multiverse with an infinite variety of possible laws and realities. Our experience is just particular to our particular reality, however broad. This suggests that there are no real absolutes and that everything is subject to change. This is consistent with Buddhism. Others have studied the problem of division in reality. The various disciplines like sociology, humanity, science, law, medicine, religion, etc. are intended to maximally reflect the particular aspects of reality they describe. However, their division has been an impediment to their natural levels of interactivity. Ken Wilber, who also makes analogies to physics, has presented the concept of the Integral Vision which attempts to integrate different fields.

What we present here is a more fundamental way to accomplish this synthesis through extending the paradigm of science with Siler's metaphorm concept. The idea has merit because of the singular success of science in describing reality and empowering humanity with control of his environment. In short, our argument is that there is a singular approach to unification of the various disciplines. The success in this presentation of uniting information and physics creates the evidence for further developments. Some common ideas are explored here. We essentially present a spectrum of consciousness in terms of intelligence which is based on the variation of the frequency of information. Wilber has also discussed the spectrum of consciousness, including the existence of an electromagnetic-like spectrum. There is commonality in this idea to the ideas presented in this work. We have directly presented that information behaves like the electromagnetic field, which we call the information field. To that end, we might say that these ideas extend the work of Wilber to a more detailed comparison with physics. We note that it has occurred in this work that both God and thought energy are imaginary components of consciousness. Thus, the source of creation is God or infinite thought. When thought becomes a representation it produces real energy. Energy is essentially vibration. Thus, all of physical reality is vibration. The inability of science to resolve the nature of the mind-body duality is equivalent to the inability of science to resolve the existence of God. Nonetheless, both God and the mind form the true source of reality.

INDEX

A

action   5, 17, 30-1, 48-50, 58-9, 66-7, 74, 86
alpha 58
angular 35-7, 39
anti-themes 2, 6, 8-10
AOT 74-5
ATO 74
attachment 3-4, 24
autistic 52, 61-2

B

Bacon 16, 73
bad 5-6, 27, 66, 73, 112
beta 58
Big Bang 79, 95-6
binary 35, 38, 41, 65, 67-9
bit   13, 26, 41, 44-5, 61, 68, 73, 84, 96
Bohm 72, 80
Bohr 72, 76
brain   ix, 8-15, 23-4, 27-8, 30, 47, 58, 62, 64, 69-72, 75, 79, 99, 111
brainwave 58, 62, 111
Buddhism   1-2, 4-6, 24, 27, 65-9, 71-2, 88, 95, 106-7, 113

C

capacitance 24
certainty 68, 86-8, 103, 110
Christianity   1-2, 5-6, 95, 106
clock   10-11, 32-3, 41-2, 93, 104, 110
communication   ix, 16, 32, 36, 38-40, 42, 47, 78, 80, 86-7, 90-1, 94, 97-9, 102, 104, 110-11
complex   2, 11-12, 21, 29, 53, 58, 66, 71, 79, 87, 90, 95-6, 100-1, 104, 109-11
complimentarity   14, 76-7, 86, 98
concave 18
concentration 11-12, 14, 18, 66
concept   vii, ix-x, 5-7, 10, 19, 21, 30, 42, 46-7, 57-8, 60, 63-6, 73-6, 88-90, 107, 111-14
consciousness   vii, 18, 24, 28, 46-7, 51, 58, 63, 105, 110, 112, 114
continuity   2, 16-17, 27, 48, 95, 102, 106, 110
convex 18
current   7, 17, 21-6, 33-4, 38, 40, 42, 48, 55, 84, 102, 107, 109-10

V

velocity   35-6, 45-6, 59, 90, 93, 100, 104, 110-11
voltage   ix, 21-3, 25, 31, 42, 103, 107

W

Wave   19, 23, 41, 54, 58, 61, 86
wave function   41, 54, 61
weak force   18
wheel   35-40
Wheeler   61, 73, 82, 84
Wilber   vii, 3, 89, 113-14
will   1-3, 5, 12-14, 27-9, 44-6, 48-9, 53-6, 59-62, 65-7, 72-3, 77-9, 90, 97, 103-4, 106, 109-10
Wittgenstein   72, 78

Y

yang   5, 18-20, 23-4, 26-8, 68, 95, 107
yin   5, 18-20, 23-4, 26-8, 68, 95, 107

www.ingramcontent.com/pod-product-compliance
Lightning Source LLC
Chambersburg PA
CBHW022005170526
45157CB00003B/1148